Lecture Notes in Earth Sciences 81

Editors:
S. Bhattacharji, Brooklyn
G. M. Friedman, Brooklyn and Troy
H. J. Neugebauer, Bonn
A. Seilacher, Tuebingen and Yale

D1104381

Springer
Berlin
Heidelberg
New York
Barcelona
Hong Kong
London
Milan
Paris
Singapore
Tokyo

Yigang Song   German Müller

Sediment-Water Interactions in Anoxic Freshwater Sediments

Mobility of Heavy Metals and Nutrients

With 56 Figures and 10 Tables

 Springer

Authors

Dr. Yigang Song
Poly Research Corp.
125 Corporate Drive, Holtsville, NY 11742, USA

Prof. Dr. Dr. h.c. mult. German Müller
Institute of Environmental Research, University of Heidelberg
Im Neuenheimer Feld 236, D-69120 Heidelberg, Germany

"For all Lecture Notes in Earth Sciences published till now please see final pages of
the book"

Cataloging-in-Publication data applied for

Die Deutsche Bibliothek - CIP-Einheitsaufnahme

Song, Yigang:
Sediment water interactions in anoxic freshwater sediments : mobility
of heavy metals and nutrients ; with 10 tables / Yigang Song ;
German Müller. - Berlin ; Heidelberg ; New York ; Barcelona ;
Budapest ; Hong Kong ; London ; Milan ; Paris ; Singapore ; Tokyo :
Springer, 1999
   (Lecture notes in earth sciences ; 81)
   ISBN 3-540-65022-9

ISSN 0930-0317
ISBN 3-540-65022-9 Springer-Verlag Berlin Heidelberg New York

© Springer-Verlag Berlin Heidelberg 1999
Printed in Germany

The use of general descriptive names, registered names, trademarks, etc. in this
publication does not imply, even in the absence of a specific statement, that such
names are exempt from the relevant protective laws and regulations and therefore
free for general use.

Typesetting: Camera ready by authors
SPIN: 10492801   32/3142-543210 - Printed on acid-free paper

Contents

1 Abstract

Major rivers within Germany (Rhine, Neckar, Main, Weser, and Elbe) drain densely populated and important industrial areas. The rivers had been polluted with heavy metals and organic pollutants by industrial and municipal emissions mainly before 1970 or - in the case of the Elbe - before 1990.

An important ecological problem still exists in the high heavy metal concentration of the river sediments. To assess the risk of heavy metal remobilization from sediments, porewater and sediment samples were examined from 10 sites in these rivers. For comparison porewaters from Lake Constance were also analyzed. Further investigations were carried out on sediment profiles of the Lean River (China) draining a hinterland with one of the largest copper mines of the world, and the Oka River (with its important tributary, the Moscow River) in Russia. The distributions of Fe, Mn, Cd, Zn, Pb, Cu, Cr, and Co in the porewater and sediments are reported. In addition, NO_3^-, SO_4^{2-}, NH_4^+, PO_4^{3-}, Ca^{2+}, Mg^{2+}, Br^-, alkalinity and pH were determined in the porewaters to study early diagenetic processes.

Generally, the depth profiles of NO_3^-, Mn^{2+}, Fe^{2+}, and SO_4^{2-} in the sediments are similar to those reported in other organic-rich sediments: the concentrations of NO_3^- and SO_4^{2-} decrease with depth, while the concentrations of Mn^{2+} and Fe^{2+} increase. This is related to the mineralization of organic matter. NO_3^-, Mn oxide, Fe oxide, and SO_4^{2-} are subsequently reduced during degradation of organic matter. Furthermore, these reactions occur directly below the sediment-water interface (between 0-20 cm), suggesting strong anoxic condition in the sediments.

Rates of NO_3^- reduction appear to depend on temperature. The higher the temperature, the faster the reaction rate. As the mineralization of organic matter is mainly biologically catalyzed, higher temperature may result in higher bacteria activities, and consequently higher degradation rate of the organic matter. In addition, the availability of labile organic matter and NO_3^- has also an effect on the reaction rates.

During the mineralization of organic matter, Mn^{2+} and Fe^{2+} are released into the porewater as a result of the reduction of Mn and Fe oxides. Solubility calculations indicate that the porewater is supersaturated in respect to rhodochrosite ($MnCO_3$) and siderite ($FeCO_3$). Mn^{2+} and Fe^{2+} seem to be controlled by the formation of Mn and Fe carbonate in the anoxic porewater. Mn^{2+} and Fe^{2+} diffuse upward due to their concentration gradients. They are reoxidized and precipitated as Mn and Fe oxides in the oxic surface layer. However, significant accumulations of particulate Mn and Fe in the surface sediments have not been found. This can be attributed to high geogenic Fe and Mn concentrations already existing in the sediments. In addition, mixing processes of the sediments by bioturbation and/or resuspension might prevent an enrichment.

As products of the mineralization of organic matter, the concentrations of NH_4^+ and alkalinity increase with depth. They are very different at different sites due to different intensity of organic matter decomposition. In most cases, NH_4^+ and alkalinity profiles can be separated into two zones: the SO_4^{2-} reduction zone and the CH_4 fermentation zone.

PO_4^{3-} is released into porewater not only by the degradation of organic matter, but also by the reduction of Fe oxides, at which PO_4^{3-} is adsorbed. In anoxic sediments, PO_4^{3-} concentrations appear to be controlled by the formation of vivianite $(Fe_3[PO_4]_2 \cdot 8H_2O)$. The porewater is supersaturated in respect to vivianite, suggesting the slow kinetics of vivianite precipitation.

The concentrations of Br^- increase with depth. The very high positive correlations between Br^-, NH_4^+, and alkalinity reflect that bromine, originally a constituent of the organic matter in the sediments, is released as Br^- during the decomposition of organic matter. Therefore, the sediments act both as sinks and sources for bromine in aquatic systems.

The high concentrations of NH_4^+ and PO_4^{3-} in the porewater result from the mineralization of organic matter. However, the flux of NH_4^+ and PO_4^{3-} from the sediments into overlying water is low. This can be explained by the existence of an oxic surface layer. In this layer, PO_4^{3-} diffused from the deeper sediments can be adsorbed on the freshly formed Fe oxides, while NH_4^+ will be oxidized to NO_3^-. When the oxic layer is destroyed (e.g. by flood or dredging), a release of NH_4^+, PO_4^{3-}, and Fe^{2+} from the sediments into the overlying water may occur. The spontaneous oxidation of NH_4^+ and Fe^{2+} can cause a severe O_2-depletion in the overlying water, which might lead to fish kills. In addition, NH_4^+ is toxic for fish, too.

As compared with the supernating water, the concentrations of the metals Cu, Pb, Zn, and Cd in the porewaters of the anoxic sediments are considerably lower. This can be explained by the formation of highly insoluble metal-sulfides. In contrast, Cr concentrations in the anoxic porewater are generally higher than in the supernating water. The explanation lies in the fact, that Cr does not form Cr-sulfide and therefore its solubility is not determined by the HS^- concentration of the porewater.

The higher concentrations of Co in the porewater find their explanation in the relatively higher solubility of CoS as compared with other heavy metals. With the reduction and dissolution of Fe and Mn oxides, Cr and Co are simultaneously released into the porewater.

In the Neckar River sediments, the Acid Neutralizing Capacity (ANC) of the sediments is much higher than the Acid Producing Capacity (APC). This is due to the high carbonate content (6.5 % - 25 %) of the sediments. Therefore, an oxidation of the anoxic sediments must not necessarily lead to an acidification. In this case, a significant release of heavy metals by the resuspension of the sediments cannot be expected. Considering seasonal variations of the porewater profiles, the peaks of dissolved Cd, Zn, Pb, and Cu at the sediment-water interface may not be caused by sediment leaching, but may result from the decomposition of organic matter containing these metals.

In summary, the porewater profiles show that heavy metals are not leached from but rather diffuse into the sediments. The sediments therefore act as a sink rather than a source.

2 Introduction

Pollutants discharged into aquatic systems are mostly adsorbed on suspended particles and finally accumulate in sediments. As many aquatic organisms spend a major portion of their life in or on sediments, polluted sediments provide a pathway for the chemicals to food chain organisms, and finally to humans.

Numerous studies have shown that sediment-water interactions in natural aquatic systems play an important role in controlling transport processes of pollutants (Eck and Smits 1986; Gobeil et al. 1987; Morfett et al. 1988; Jahnke et al. 1989; Gerringa 1990; Ivert 1990; Carignan and Lean 1991; Dahmke et al. 1991; Barbanti et al. 1992a; Williams 1992).

The aim of this work is to understand the relationships between mineralization of organic matter and the mobility of heavy metals and nutrients in the heavily polluted sediments. The study includes two parts: the first focuses on describing the diagenetic processes, involving decomposition of organic matter and recycling of nutrients (NO_3^-, NH_4^+, PO_4^{3-}, and alkalinity); the second emphasizes the associated chemical behavior of heavy metals in the contaminated sediments, and their mobility during the early diagenetic processes.

The results should enable us to know whether heavy metals and nutrients are released into the overlying water under changes of physicochemical conditions. This is important to water management and sediment cleanup plans.

2.1 Early diagenesis

In natural water, sediment-water interaction plays a fundamental role in biogeochemical cycling of elements. One of the important reactions is the mineralization of organic matter during early diagenesis, which is mainly biologically catalyzed. This process leads to a change in concentration, temporal and spatial distribution, and speciation of elements in water, suspended particles, and sediments. Therefore, the knowledge about this process is important to evaluate the chemical behavior of heavy metals in sediments.

Organic matter in sediments mainly derives from phytoplankton, zooplankton, and other organic materials. The commonly accepted model for decomposition of organic matter is illustrated in Fig. 2.1. Organic matter is oxidized by the oxidant yielding the greatest free energy change per mole of organic carbon oxidized. When this oxidant is depleted, oxidation will proceed utilizing the next most efficient oxidant (Froelich et al. 1979; Berner 1980):

1. $(CH_2O)_{106}(NH_3)_{16}(H_3PO_4) + 138\ O_2$
$$\rightarrow 106\ CO_2 + 16\ HNO_3 + H_3PO_4 + 122\ H_2O$$
$$\Delta G° = -3190\ kJ/mole$$

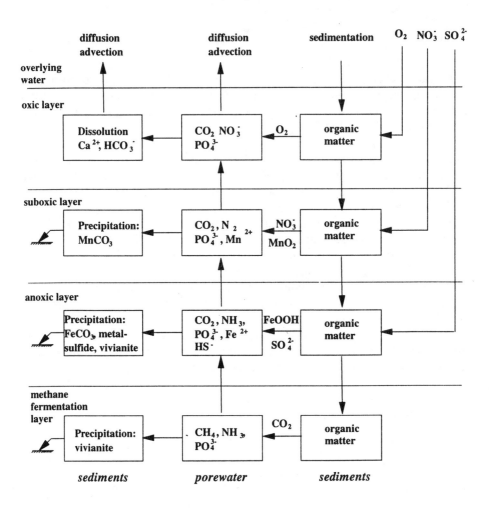

Fig. 2.1. Schematic view of the early diagenetic processes

2. $(CH_2O)_{106}(NH_3)_{16}(H_3PO_4) + 236\ MnO_2 + 472\ H^+$
$$\rightarrow 236\ Mn^{2+} + 106\ CO_2 + 8\ N_2 + H_3PO_4 + 366{,}H_2O$$
$$\Delta G^\circ = -3090\ kJ/mole$$

3. $(CH_2O)_{106}(NH_3)_{16}(H_3PO_4) + 94.4\ HNO_3$
$$\rightarrow 106\ CO_2 + 55.2\ N_2 + H_3PO_4 + 177.2\ H_2O$$
$$\Delta G^\circ = -3030\ kJ/mole$$

4. $(CH_2O)_{106}(NH_3)_{16}(H_3PO_4) + 424\ FeOOH + 848\ H^+$
$$\rightarrow 424\ Fe^{2+} + 106CO_2 + 16\ NH_3 + H_3PO_4 + 742\ H_2O$$
$$\Delta G^\circ = -1330\ kJ/mole$$

5. $(CH_2O)_{106}(NH_3)_{16}(H_3PO_4) + 53\ SO_4^{2-}$
$$\rightarrow 106\ CO_2 + 16\ NH_3 + 53\ S^{2-} + H_3PO_4 + 106\ H_2O$$
$$\Delta G^\circ = -380\ kJ/mole$$

6. $(CH_2O)_{106}(NH_3)_{16}(H_3PO_4) \rightarrow 53\ CO_2 + 53CH_4 + 16\ NH_3 + H_3PO_4$
$$\Delta G° = -350\ kJ/mole$$

In the oxic layer, reaction 1 continues as long as sufficient O_2 is available for the oxidation process. In the suboxic layer, MnO_2 reduction (reaction 2) and NO_3^- reduction (reaction 3) occur simultaneously. After consumption of NO_3^- and reactive MnO_2, oxidation is driven by Fe oxide reduction (reaction 4) and SO_4^{2-} reduction (reaction 5) in the anoxic layer, and finally by methane fermentation (reaction 6).

The knowledge of early diagenetic processes in sediments is mostly obtained from marine sediments within the framework of global element cycling (Bender et al. 1977; Froelich et al. 1979; Aller 1980a,b; Berner 1980; Elderfield et al. 1981; Jahnke et al. 1982; Balzer 1984; Pettersson and Boström 1986; Westerlund et al. 1986; Jahnke et al 1989; McCorkle and Klinkhammer 1990; Shaw et al. 1990; Bruland et al. 1991; Dahmke et al. 1991; Boudreau et al. 1992; Wu et al. 1992). As the chemical environments in freshwater and marine systems usually differ, varied redox reactions may dominate the decomposition of organic matter in sediments. For example, quantitative differences in the input of organic matter between freshwater and marine sediments likely affect the diagenetic processes. The depth of O_2 penetration depends on its downward diffusion from overlying water and consumption of O_2 by aerobic decomposition of organic matter. Freshwater sediments generally have higher organic matter contents than marine sediments. This leads to a rapid consumption of O_2. As a result, O_2 is depleted below a few millimeters of the sediment-water interface in freshwater sediments (Jørgensen 1983). In deep sea sediments where the organic input is low, O_2 may diffuse downward to several centimeters, and aerobic decomposition is the main process (Murray and Grundmanis 1980; Billen 1982; Reimers and Smith 1986; Jahnke et al. 1989).

In marine and most freshwater sediments, NO_3^- reduction is not a dominating process because of its low concentration in overlying water and porewater (Bender and Heggie 1984; Balzer 1989). In contrast, rivers draining agricultural land commonly have high NO_3^- concentrations, and denitrification may be an important process (Jørgensen and Sørensen 1985; Sagemann et al. 1994).

Another significant difference is the high concentration of SO_4^{2-} in porewater of the marine sediments. Compared to methane bacteria, SO_4^{2-} reducers have a higher affinity for H_2, acetate, methanol, which can be used for SO_4^{2-} reduction and methane fermentation. Therefore, active methane fermentation occurs only after the consumption of SO_4^{2-} and is spatially separated from areas of sulfate reduction (Winfrey and Zeikus 1977; Jørgensen 1983). The difference of sulfate availability between marine and freshwater sediments results in different anoxic decomposition processes of organic matter. High concentrations of SO_4^{2-} (20-30 mM) in marine sediments are responsible for most of the anoxic oxidation (Capone and Kiene 1988; Balzer 1989). In freshwater sediments SO_4^{2-} (about 0.2 mM) may be depleted rapidly within the sediments because of the oxidation of organic matter and methane fermentation becomes the dominating process (Mountfort and Asher 1981; Kuivila et al. 1989).

Seasonal temperature changes can bring about appreciable variation of biological activities in shallow water sediments, and consequently have an influence on the diagenetic processes, which is mainly biologically catalyzed (Billen 1982; Holdren and Armstrong 1986). Jørgensen and Sørensen (1985) reported that the temperature coefficient (Q_{10}, the factor of reaction rate increase per 10°C temperature increase) is 2.7 for O_2 oxidation and 2.1 for SO_4^{2-} reduction in the sediments of Norsminde Fjord, Denmark. In a laboratory experiment, Sagemann et al. (1994) found a significant correlation between denitrification rate and temperature within the sediment from Weser Estuary, Germany.

2.2 Behaviour of heavy metals and nutrients during early diagenesis

In natural waters, heavy metals and nutrients are mainly adsorbed and/or bound to particles such as Fe/Mn oxides, clay minerals and organic materials. The mineralization of organic matter leads to a change of the Eh-pH conditions and the chemical composition of the sediments. This results in a series of new chemical equilibria between sediments and water. Adsorption/desorption on Fe/Mn oxides, precipitation/dissolution of minerals, and complexation with organic and inorganic colloids are the predominant processes. These reactions mainly control the cycling of elements in aquatic systems.

Previous studies have shown that Fe/Mn oxides provide important adsorbing surfaces and constitute significant sinks for heavy metals and nutrients in surface water (Tessier et al. 1985; Sigg 1986; Sigg et al. 1987). In the anoxic sediment layer, heavy metals and phosphates bound to the oxides can be released into porewater, following the reduction of Fe/Mn oxides. On the other hand, Fe^{2+} and Mn^{2+} diffusing into the oxic sediment layer will be oxidized to Fe/Mn oxides and immobilized. The freshly formed Fe/Mn oxides are very efficient scavengers for heavy metals and phosphate in this layer. Therefore, the cycling of Fe and Mn may play an important role in transport processes of heavy metals and phosphorus (Davison et al. 1982; Pedersen and Price 1982; Hamilton-Taylor et al. 1984; Gendron et al 1986; Salomons et al. 1987; Francis and Dodge 1990; McCorkle and Klinkhammer 1990; Dahmke et al. 1991; Balistrieri et al. 1992; Johnson et al. 1992; Morse and Arakaki 1993; Williams 1992; Matsunaga et al. 1993).

Another important reaction is the formation of minerals during the early diagenetic processes. Precipitation/dissolution of calcite, dolomite, siderite, rhodochrosite may influence porewater concentrations of Ca^{2+}, Mg^{2+}, Fe^{2+}, Mn^{2+}, and alkalinity (Emerson 1976; Matisoff et al. 1981; Postma 1981; Kuivila and Murray 1984; Dahmke et al. 1986; Norton 1989; Wallmann 1990; Vuynovich 1989; Dahmke et al 1991). Numerous studies have attested that the formation of phosphorus minerals (e.g. vivianite, reddingite) may control phosphate concentrations in anoxic porewater (Nriagu and Dell 1974; Emerson 1976; Suess 1979; Elderfield et al. 1981; Postma 1981). In the last few years, the role of sediment sulfides in controlling the distribution of heavy metals between sediments and porewater has been demonstrated. HS^- is produced due to SO_4^{2-} reduction in anoxic sediments. It can form highly insoluble metal sulfide precipitates with metal ions (Lee and Kittrick 1984a,b; Carignan and Nriagu 1985; Carignan and Tessier 1985; Gendron et al. 1986; Wallmann 1990).

It is noticeable that many precipitation/dissolution reactions are kinetically inhibited. Supersaturation of minerals is often observed in porewater before precipitation of solid phase actually occurs (Nriagu and Dell 1974; Postma 1981; Holdren and Armstrong 1986). High temperature, presence of nucleating surfaces, and biological activities can dramatically enhance the kinetics (Morel and Hering 1993). In addition, the existence of organics and organic colloids may lead to a supersaturation of sulfides (Boulegue et al. 1982).

Under changes of physicochemical conditions (e.g. Eh-pH), the equilibrium of precipitation/dissolution is disturbed. This leads to a redistribution of heavy metals and phosphorus between sediments and porewater, and can influence the mobility of these components (Kersten et al. 1985; Wallmann 1990; Gambrell et al. 1991; Calmano et al. 1992).

The cycling of heavy metals and nutrients in sediments is controlled by precipitation/dissolution of minerals, but the mobility and bioavailability may be related to the organic and inorganic complexation. Porewater studies reflect that heavy metal concentrations are significantly higher than could be predicted from metal-sulfide solubility (Elderfield et al. 1981; Carignan and Nriagu 1985). In laboratory experiments, Salomons et al. (1987) and Wallmann (1992b) reported that inorganic complexes (e.g. with sulfide and chloride) can explain the supersaturation of metal-sulfides. Organic complexing materials such as humic substances are important because of their high affinity to heavy metals (Elderfield 1981; van den Berg and Dharmvanij 1984; Douglas et al. 1986). Previous studies show that organic colloids (e.g. humic substances) increase with depth as a result of the degradation of organic matter (Orem et al. 1986; Chin and Gschwend 1991). This may enhance the mobility and bioavailability of heavy metals during the early diagenesis (Elderfield 1981; Douglas et al. 1986).

2.3 Methods used in the study of early diagenesis

Until now, three kinds of studies have been carried out to study diagenetic processes. The first are laboratory experiments (core incubation). For this method, sediment cores are brought into the laboratory, where water and biological debris (e.g. plant, shrimps) are added on the top of the sediments. The release of nutrients, dissolved organic carbon (DOC), and heavy metals is then measured in the water column. This method provides a direct measurement of the degradation processes under changes of physicochemical conditions. It is relatively simple and inexpensive, and can be conducted with a wide variety of parameters such as temperature, input of organic matter, redox potential, chemical composition of the water (Matisoff et al. 1981; Gerringa 1990; Wu et al. 1992; Søndergaard et al. 1992; Matsunaga et al. 1993; Morse and Arakaki 1993; Sagemann et al. 1994). Radiotracers are often used to study redox reaction mechanisms (Winfrey and Zeikus 1977; Gunnarsson and Rönnow

1982; Fossing and Jørgensen 1990; Elsgaard and Jørgensen 1992, Lee and Fisher 1992). The disadvantage of this method is the unknown relationship between the experiments and the reality in nature. In addition, it does not provide relationships between the sediments and the porewater.

The second is the use of sediment traps (flux chambers), which are pushed some centimeters into sediments or above sediments. This method allows to study the role of suspended particles depositing on sediments. It gives a direct and real in situ measurement for the recycling of nutrients and heavy metals from sediments. Since the chambers are inserted on site, this method does not substantially alter the studied natural system, whereas a disturbance generally occurs with the core incubation method. The disadvantage of this method is that it does not provide information on the changes in different sediment layer and the distributions of elements between sediments and porewater (Aller 1980a,b; Davison et al. 1982; Hamilton-Taylor et al. 1984; Sakata 1985; Sigg 1986; Westerlund et al. 1986; Sigg et al. 1987; Barbanti et al. 1992a,b)

The third method consists of the analysis of porewaters and sediments. Since any change in the diagenetic processes should be reflected by concentration gradients with depth, this study gives a relationship between sediments and porewater. Assuming a steady state, a diagenetic model can be used to evaluate the decomposition rate of organic matter and the role of different oxidants. The diffuse flux from (into) the sediments can be calculated by Fick's First Law based on the concentration gradients. The disadvantage of this method is that it does not offer any information about the reactions occurring directly on the sediment surface and the role of freshly depositing particles. In addition, porewater sampling requires more equipment to prevent the oxidation of samples during sample collection.

Recently, porewater sampling by in situ dialysis has been widely used (Kuivila and Murray 1984; Carignan and Nriagu 1985; Carignan 1985; Kuivila et al. 1989; Vuynovich 1989; Carignan and Lean 1991; Dahmke et al. 1991). Compared to the conventional squeezing or centrifuging methods, the use of dialysis samplers ('Peeper') requires less equipment and does not disturb the primary conditions in the porewater.

3 Methods and materials

3.1 Location of the study area

3.1.1 Neckar River and its tributaries

The Neckar River is a tributary of the Rhine; it drains densely populated and important industrial areas in Southwest Germany (Fig. 3.1). Before 1973, The Neckar River was heavily polluted by industrial and municipal discharges, which is well documented by water, biota, and sediment research (Förstner and Müller 1973; Reinhard and Förstner 1976; Müller and Prosi 1977; Müller and Prosi 1978; Müller 1980, 1981, 1986; Müller et al. 1993).

There are 27 locks along the navigable middle and lower section of the Neckar River. Large amounts of suspended particles deposit in front of the dams due to low water flow rate. At site Lauffen alone 700,000 m^3 of heavily polluted sediments accumulated under overlying water. These contaminated sediments pose problems for this aquatic ecosystem.

The sampling sites represent different depositional environments of the Neckar River and its tributaries (Fig. 3.1).

1. Neckar at Lauffen. This site was chosen because of the large amounts of contaminated sediments and the high heavy metal concentrations in the sediments. It is a representative site for the ecological problems in the Neckar River. The river flows very slowly at this site, which leads to a deposition of fine particles and biological debris. This gives good preconditions for the research of diagenetic processes.

2. Neckar at Kochendorf. The sediments were moderately polluted. The carbonate contents was highest at this site (about 30 %; Müller et al. 1993).

3. Neckar at Wieblingen. The sediments were moderately polluted. The dam is often open because of large water flow after heavy rains. As a result, the sediments are removed and transported to a lower section of the river.

4. The Elsenz River at Neckargemünd. The Elsenz is a tributary of the Neckar River. It drains mainly agricultural area. The sediments mainly derive from soil runoff and have low contents of heavy metals.

5. The Enz River at Vaihingen. The Enz is one of the largest, and most heavily polluted tributaries. This river has great influence on the Neckar River. The level of biological productivity is high and is reflected by the surface organic carbon (7.1 %). Carbonate content was about 6 % (Müller et al. 1993).

6. The Schwarzbach River at Neckarbischofsheim. The Schwarzbach is a tributary of the Elsenz River. The water flow is low. The sampling site is located at Neckarbischofsheim, where serious fish kills occurred in April 1992.

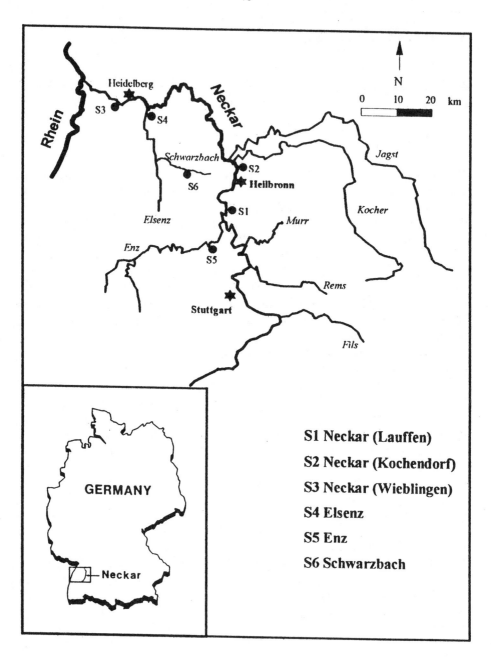

Fig. 3.1. The sampling sites in the Neckar River and its tributaries

3.1.2 Major rivers and Lake Constance in Germany

Systematic studies have shown that most of the major rivers in Germany were heavily polluted by heavy metals - Rhine, Neckar, Main, Weser, and Elbe (Förstner and Müller 1973). In spite of large improvement since 1970, there is significant enrichment in the sediments, especially in 'old' sediment layers.

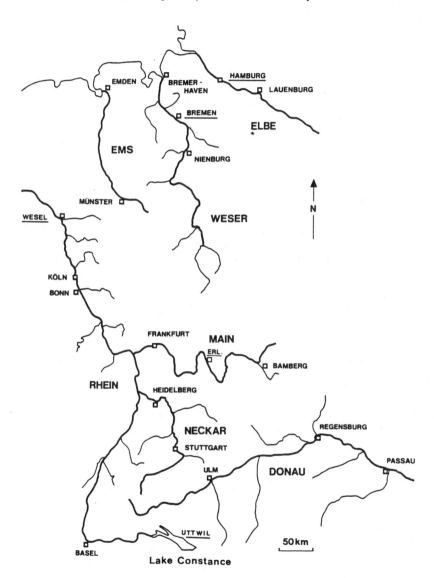

Fig. 3.2. Sampling sites in the major rivers and Lake Constance in Germany.

Table 3.1. Sampling sites

sampling site	sampling data	water depth (m)
Elbe (Hamburg)	Dec. 1. 1988	3.40
Weser (Bremen)	Dec. 1. 1988	3.40
Rhine (Wesel)	Sep. 23. 1988	4.50
Main (Erlabrunn)	Nov. 11. 1988	2.20
Neckar (Lauffen)	Oct. 9. 1988	3.20
Lake Constance (Uttwil)	July. 1988	15.0
Neckar (Lauffen)	Oct. 1. 1991-June. 15. 1993	1.50
Neckar (Kochendorf)	June 15. 1992	1.40
Neckar (Wieblingen)	Nov. 15. 1991	2.10
Elsenz (Neckargemünd)	Jan. 10. 1992	1.40
Enz (Vaihingen)	June 15. 1992	1.50
Schwarzbach (Neckarbischofsheim)	Nov. 10. 1992	0.80
Lean River (China)	June 1993	2.50
Moscow River (Russia)	July 24. 1994	0.60
Oka River (Russia)	July 24. 1994	0.60

Lake Constance is the second largest lake in central Europe (surface 540 Km2, max. depth 252m). It receives the main water and particulate input from the Rhine River (Alpenrhein). In contrast to the major rivers in Germany, Lake Constance was slightly contaminated by heavy metals.

Data on the sampling sites are given in Table 3.1 and Fig. 3.2.

3.1.3 Lean River (China), Moscow River and Oka River (Russia)

Lean River (279 km long) is situated near the largest opencast mine in South China (Fig. 3.3). Previous investigations revealed that sediments of the Lean River are contaminated with Cu, Zn, and Pb. This is mainly due to the discharge of acid waste water from the mine. In June 1993, sediment core and porewater samples were collected at Caijiawan (Fig. 3.3).

Oka River (1500 km long) is the second largest tributary of the Volga River. The Moscow River, which is polluted by heavy metals, flows into the Oka River at Kolomna. In July 1994, Two peepers were installed in the Oka River and Moscow River at Kolomna (Fig. 3.4).

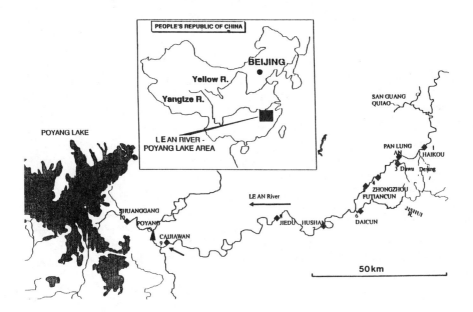

Fig. 3.3. Location of core and porewater collected from Lean River (China)

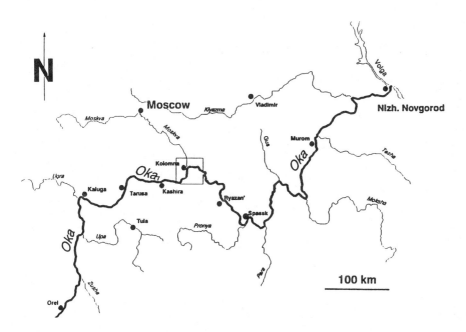

Fig. 3.4. Sampling sites in the Oka River and Moscow River

3.2 Sampling

Sediment cores were taken with a piston corer and sectioned every 2 cm within 24 hours after collection. Porewater samples were collected by in situ dialysis at the same sites between October 1991 and June 1993. The dialyzer used in this study was similar in design to that described by Hesslein ('Peeper', 1976). It contains two vertical rows of 65 mm long x 6 mm wide x 10 mm deep horizontal chambers with a 1 cm spacing (Fig. 3.2). The 'Peeper' was filled with distilled water and covered with a biologically inert polycarbonate membrane. Before insertion, the 'Peeper' was deaerated in distilled water by a nitrogen-treatment. After an equilibration time of two weeks, the 'Peeper' was taken out, and the water samples were immediately removed with plastic syringes (which were washed with 2 N HNO_3) in a nitrogen atmosphere. The porewater samples were then divided into three parts: one was acidified with 20 µl HNO_3 and stored at 4 °C for metal analysis; one was stored at 4 °C for anion analysis; pH, alkalinity, and conductivity were measured immediately in the third part of the samples.

3.3 Analysis.

3.3.1 Solid phases

The water content of the sediments was determined by drying the samples to constant weight at 60 °C. Porosity was calculated from water content and particle density.

Organic carbon (C_{org}) and sulfur were measured with a LECO CS-Analyzer. Before the analysis, the samples were extracted with 1 N HCl to remove inorganic carbon.

Total phosphorus was determined by spectrophotometry at 436 nm wavelength, following a digestion with a HNO_3-$HClO_4$ mixture.

The carbonate content was determined using the 'Karbonat-Bombe' designed by Müller and Gastner (1971). 5 ml of 20 % HCl was added to 0.7 g of dried sediment in a glas container, after closing the instrument and shaking it, a CO_2 pressure proportional to the carbonate content of the sample is built up.

The fraction (< 20 µm) of the sediment samples was digested with HCl-HNO_3 (aqua regia). Fe, Mn, Cd, Zn, Pb, Cu, Cr, and Co were analyzed with a Perkin Elmer Atomic Absorption Spectrophotometer model 3030.

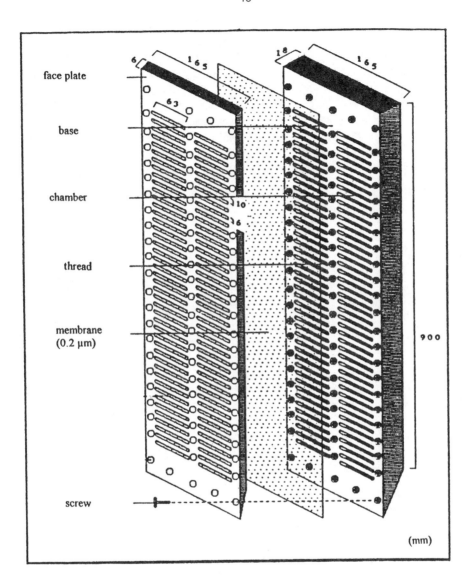

Fig. 3.5. The Peeper

3.3.2 Porewater.

pH was measured with a pH meter. Conductivity was analyzed with a WTW (LF 191) conductivity meter. Alkalinity was determined by titration with 0.036 N H_2SO_4.

NH_4^+ was measured at 655 nm wavelength by a Spectrophotometer, model Spectronic 1201.

Dissolved Cd, Zn, Pb, Cu, Cr, and Co were analyzed using a Perkin Elmer HGA 500 graphite furnace. Fe, Mn, Ca, and Mg were determined by Flame-AAS, Model Perkin Elmer 3030. The precision of the metal analyses based on replicate samples was 10 % for Fe, Mn, Cu, and Zn; 15 % for Co, Cr, Cd and Pb.

NO_3^-, SO_4^{2-}, Cl^-, Br^- were analysed by a Dionex 4000i Ionchromatograph.

4 Results and discussion

4.1 Early diagenesis in the sediments of the Neckar River and its tributaries

4.1.1 Decomposition of organic matter by different oxidants

Porewater Mn^{2+}, NO_3^-, Fe^{2+}, SO_4^{2-}, NH_4^+, alkalinity, PO_4^{3-}, pH, and conductivity profiles in the sediments at Lauffen are shown in Fig. 4.2. They demonstrate anaerobic decomposition of organic matter below the sediment-water interface.

Particles sinking to sediments include particulate organic matter (POC), ferric oxides, carbonates, and silicates (Sigg 1986). On the sediment surface, these particles are often resuspended by hydrodynamic forces or biological activities, before they finally become part of the sediment.

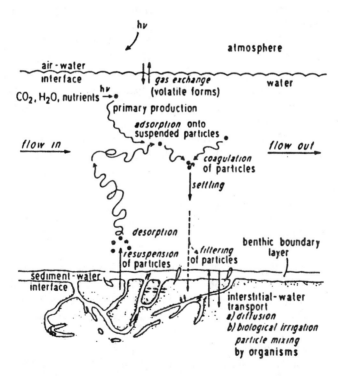

Fig. 4.1. Transport processes near the sediment-water interface (Santschi et al. 1990)

Fig. 4.1 illustrates the transport processes near the surface sediment layer. The benthic boundary layer at the sediment surface can be described as a buffer zone between sediments and water. Its thickness is related to hydrodynamic conditions and biological activities. At the sediment surface, organic matter such as algae and plant detritus is rapidly oxidized by O_2 in overlying water. The depth of O_2 penetration depends on its rate of downward diffusion, which is influenced by biological activities. In the Neckar River, minimum O_2 concentrations of the overlying water varied between 6.2-8.9 mg/l in summer, and 8.6-11.4 mg/l in winter (Deutsche Kommission zur Reinhaltung des Rheins 1991). Some aquatic fauna such as mollusks and worms were found at the surface sediments, giving the evidence of bioturbation in the surface sediments. Although O_2 penetration could extend to a few centimeters as a result of bioturbation, it was probably too low to drive an oxidation of organic matter. Therefore, a release of NO_3^- due to the oxidation of organic matter was not found in this layer.

Below the surface layer, the organic matter was subsequently oxidized by Mn oxide, NO_3^-, Fe oxide, SO_4^{2-}, and CO_2. Dissolved Mn^{2+} (Fig. 4.2a) was negligible to the depth of 4 cm, where Mn^{2+} was precipitated as MnO_2. Mn oxide was reduced and released Mn^{2+} into the porewater below this depth. NO_3^- values varied between 0.42 and 0.47 mM in the overlying water and decreased rapidly to about 0.03 mM at 20 cm depth (Fig. 4.2b). This suggests that denitrification has occurred near the sediment-water interface. Fe^{2+} was released below 8 cm depth as a result of the reduction of Fe oxide (Fig. 4.2c). Fe^{2+} and Mn^{2+} in the deeper sediments diffused into the surface layer and were reoxidized to Fe/Mn oxides by O_2. Similar porewater profiles of Fe and Mn are reported from many marine and freshwater sediments (Froelich et al. 1979; Carignan and Nriagu 1985; Eck and Smits 1986; Carignan and Lean 1991), indicating redox cycling of Fe and Mn in sediments during mineralization of organic matter.

The SO_4^{2-} concentrations ranged from 0.90 to 1.02 mM in the bottom water and decreased rapidly between 8 and 20 cm depth (Fig. 4.2d). Below this layer, SO_4^{2-} was less than 0.10 mM. The NH_4^+ values in the overlying water were 0.01-0.04 mM, but increased rapidly with depth, the maximal value was 15 mM at 78 cm depth. Alkalinity increased also with depth (Fig. 4.2ef).

This reflects that the formation of NH_4^+ and alkalinity is associated with the mineralization of organic matter in anoxic sediments (Froelich et al. 1979; Kuivila and Murray 1984; Carignan 1985; Kling et al. 1991; Sigg et al. 1991). In this process, SO_4^{2-} was the dominating oxidant. An increase of NH_4^+ and alkalinity was identified below the SO_4^{2-} reduction zone, which can only be explained by CH_4 fermentation. Jørgensen (1983) reported that the methanogenic bacteria cannot compete with the SO_4^{2-} reducing bacteria, thus the CH_4 fermentation occurs only after the consumption of SO_4^{2-} in the sediments. This was confirmed from the sediments at Lauffen. Another evidence of CH_4 production are the gas bubbles escaping from the sediments during sample collection. These gas bubbles most probably are from CH_4 production. Fe oxide and SO_4^{2-} reduction occurred near the sediment-water interface (8-20 cm depth), suggesting strong anoxic conditions in the sediments.

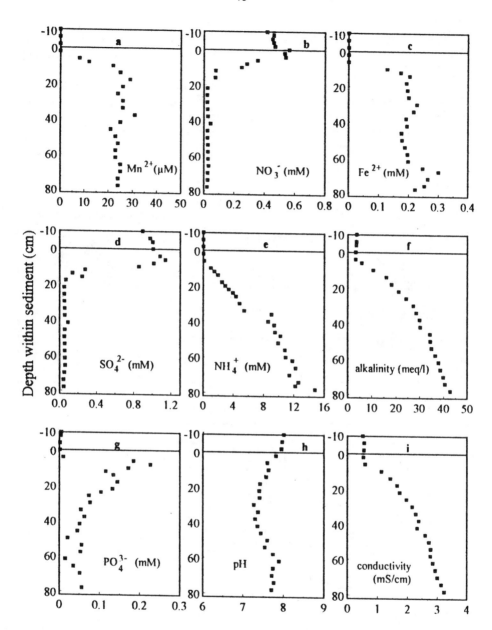

Fig. 4.2. Porewater profiles in the sediments of the Neckar River at Lauffen

The distribution of different oxidants in Fig. 4.2 displays the relative depth where they are consumed. The vertical sequence coincides with the thermodynamic principle. As the mineralization of organic matter is mainly biologically catalyzed and releases energy and nutrients for production of biomass, organisms with more energy-yielding metabolism succeed in competing with organisms with less energy-yielding metabolism (Froelich et al. 1979; Berner 1980). However, a significant boundary between oxic, suboxic, and anoxic zone described by Froelich et al. (1979) was not found in the sediments at Lauffen. The mixing of the different zones would be caused by a much higher input of organic matter compared to marine sediments.

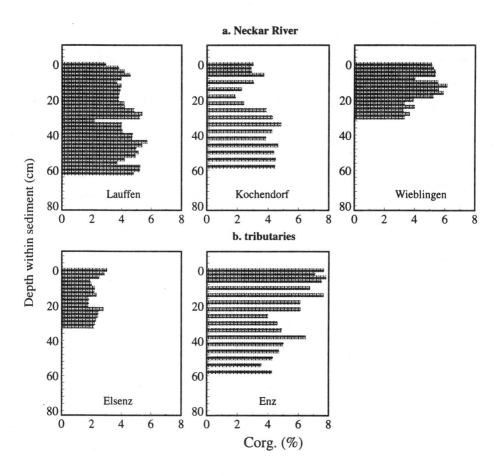

Fig.4.3. Vertical distribution of C_{org}. in the sediments of the Neckar River and its tributaries

During the mineralization of organic matter, organic P bound to biomass was also released into the porewater as PO_4^{3-} (Fig. 4.2g). In addition, inorganic P adsorbed on Fe/Mn oxide was also released with the reduction of Fe and Mn oxides. This may lead have led to an accumulation of PO_4^{3-} in the porewater. pH values varied between 7.3 and 8.0 in the porewater (Fig. 4.2h). Conductivity increased continuously with depth, indicating the release of ions into the porewater during early diagenetic processes (Fig. 4.2i).

Organic carbon (C_{org}.) contents in the sediments of the study area are plotted in Fig.4.3. In the Neckar River, the C_{org}. contents was the lowest at Kochendorf with values of 3.6 %, slightly higher at Lauffen (4.3 %), whereas the highest was at Wieblingen with 4.7 %. No significant change with depth was found in the Neckar sediments (Fig. 4.3a).

The sediments of the Elsenz River mainly accumulate from soil runoff. The average C_{org}. content was low (2.3 %) at this site. In the Enz River sediments, 7.8 % of C_{org}. was measured in the surface sediment layer, and the contents of C_{org}. decreased with depth (Fig. 4.3b).

With the decomposition of organic matter, a decrease of C_{org}. with depth in the sediments should be expected. In fact, no distinct decrease of C_{org}. with depth was found in the sediments of the Neckar River. This is in contrast with the results in marine sediments (Balzer 1989). It seems that the primary sedimentation of organic matter rather than diagenetic processes controls the contents of C_{org}. in the sediments. In the Enz River sediments, a high input of organic matter is reflected by the high C_{org}. contents in the surface sediment layer (7.8%). This may result in a rapid depletion of O_2 and an accumulation of labile organic matter in the surface sediments. Therefore, the decrease of C_{org}. with depth might be a result of the decomposition of labile organic matter within the sediments.

4.1.2 Denitrification in the sediments

At all sites, NO_3^- concentrations in the overlying water varied between 0.28 and 0.67 mM. They decreased rapidly with depth in the porewater until 20 cm depth (Fig. 4.4). Below this depth, NO_3^- concentrations ranged from 0.01 to 0.04 mM. As the aerobic oxidation of organic matter releases NO_3^- into porewater, the depth, where maximum NO_3^- concentration occurs, represents the depth of O_2 penetration. Below this depth, O_2 concentration is too low for an oxidation. Such a NO_3^- peak is reported for many marine sediments (Froelich et al. 1979; Goloway and Bender 1982; Jahnke et al. 1982; Bender and Heggie 1984; Reimers and Smith 1986; Jahnke et al. 1989). However, no peak of NO_3^- concentrations was found below the sediment-water interface in the study area. It seems that O_2 was consumed within the first millimeters below the sediment-water interface. The sampling interval (2 cm) may be too coarse to observe this peak. Similar NO_3^- profiles were also found by Kuivila and Murray (1984) in Lake Washington, and by Sagemann et al. (1994) in the Weser Estuary, Germany. Such a thin oxic layer is typical of organic-rich sediments. This feature, in conjunction with the data of Fe oxide and SO_4^{2-} reduction, indicates that anoxic conditions control porewater chemistry in the sediments of the Neckar River and its tributaries.

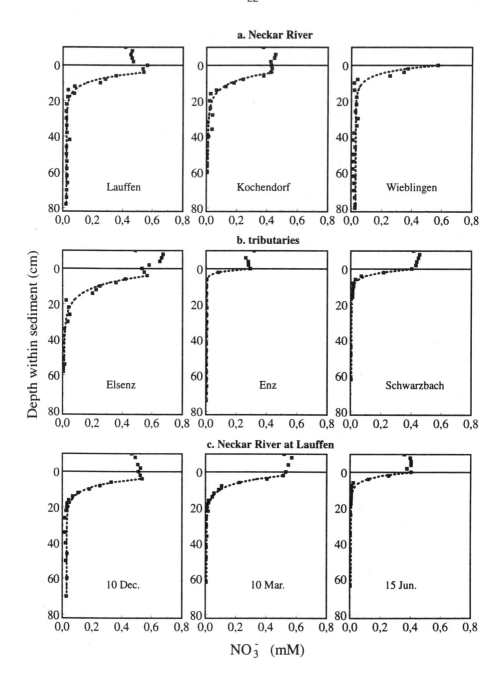

Fig. 4.4. Measured (■) and modelled (dashed linie) nitrate profiles

The stoichiometry of biomass (plankton) in marine sediments can be described as $(CH_2O)_{106}(NH_3)_{16}(H_3PO_4)$ (Froelich et al. 1979; Berner 1980; Balzer 1989). However, the ratio of elements C, N and P of the biomass in freshwater may be different from that in ocean. The stoichiometry of the biomass in freshwater is given by Sigg (1986): $(CH_2O)_{113}(NH_3)_{15}(H_3PO_4)$ based on her measurements using flux chambers. The denitrification reaction can be described by

$$(CH_2O)_{113}(NH_3)_{15}(H_3PO_4) + 99.4\ HNO_3$$
$$\rightarrow 113\ CO_2 + 57.2\ N_2 + 185.2\ H_2O + H_3PO_4$$

NO_3^- in overlying water diffuses into deeper sediments, and is reduced during the mineralization of organic matter. Under a steady state, the amount of NO_3^- diffusing into a depth x is equal to the amount of NO_3^- reduced by the organic mater ($[dC/dt]_{x=0}$; Berner 1980). Therefore, the denitrification rate in sediments can be estimated by a downward flux of NO_3^- into the sediments (Goloway and Bender 1982; Jahnke et al. 1982; Bender and Heggie, 1984; Balzer 1989):

$$[NO_3^-] = [NO_3]^\circ \exp[-\alpha\ x];$$
$$R_{NO_3} = F_{NO_3} = D_{NO_3}\ \alpha\ [NO_3]^\circ;$$

where $[NO_3^-]^\circ$: the concentrations of NO_3^- at the boundary layer,
        $[NO_3^-]$: the concentrations of NO_3^- at depth x,

        α: the fitting parameter to the measured NO_3^- profiles,
      R_{NO_3}: denitrification rate,
      F_{NO_3}: NO_3^- diffusion rate,
      D_{NO_3}: diffusion coefficient.

D_{NO_3} used in this model must be corrected by in situ temperature and porosity (Li and Gregory 1974; Berner 1980):

$$D_{NO_3} = D_{NO_3}\circ\ \phi^2\ ;$$
$$D_1\eta_1 / T_1 = D_2\eta_2 / T_2;$$

where $D_{NO_3}\ ^\circ$: free solution diffusion coefficient,
        $D_{NO_3}\ ^\circ = 1.39\ cm^2 / d$ at 18 °C. (Li and Gregory 1974),
      η: water viscosity,
      T: absolute temperature,

      ϕ: porosity.

The corrected diffusion coefficients and denitrification rates in the sediments of the study area are shown in Table 4.1. The assumption of a steady state is permissible for NO_3^-, because its concentrations and temperature in the overlying water change slowly. All measured NO_3^- profiles fit well by the model equation (Fig. 4.4), indicating the correct assumptions used in this model.

Table. 4.1. Model-estimated parameters fitting the observed NO_3^--profiles

	x*	T	ϕ	D_{NO_3}	$[NO_3^-]^\circ$	α	r*	R_{NO_3}
	cm	°C		cm^2/d	mM	1/cm		$mmol/m^2$ d
Neckar River								
Lauffen (15. Oct)	4	18	0.75	0.782	0.52	0.229	0.9833	0.94
Kochendorf	4	20	0.79	0.912	0.43	0.179	0.9847	0.70
Wieblingen	0	18	0.77	0.759	0.54	0.255	0.9706	1.0
tributaries								
Elsenz	4	8	0.72	0.547	0.56	0.153	0.9873	0.47
Enz	0	20	0.87	1.104	0.30	0.724	0.9847	2.3
Schwarzbach	0	10	0.72	0.581	0.40	0.364	0.9956	0.85
Lauffen								
winter (10. Dec)	2	14	0.75	0.704	0.49	0.226	0.9752	0.78
spring (10. Mar)	2	10	0.75	0.630	0.51	0.217	0.9467	0.69
summer (15. Jun)	0	20	0.75	0.821	0.40	0.341	0.9822	1.1

* x: the depth where denitrification began

** r: correlation coefficients

In the sediments of the study area, the estimated values for the denitrification rate ranged from 0.47 to 2.3 mmol/m^2 d. Lower denitrification rates (0.002-0.08 mmol/m^2 d) were reported by Goloway and Bender (1982) for the eastern equatorial Pacific Ocean. The NO_3^- concentrations in the sea water ranged from 0.04 to 0.06 mM, which were 5-10 fold lower compared to the Neckar River (0.28-0.67mM). Therefore, the high denitrification rate in the study area can be explained as a result of the high NO_3^- input into the sediments. This interpretation is supported by the result from Jørgensen and Sørensen (1985). They found that the denitrification rates are mostly a function of NO_3^- availability in Norsminde Fjord, Denmark.

The highest denitrification rate was measured in the Enz River. As no significant difference of NO_3^- concentrations was found between all sites, it seems that another factor rather than the NO_3^- concentrations affects the denitrification processes in the sediments. In the Enz River, high contents of $C_{org.}$ in the sediments reflect high input of biological materials at this site, which could lead to a rapid depletion of O_2 and NO_3^-. This explanation is in agreement with the observations from Goloway and Bender (1982), and Schulz et al. (1994) in marine sediments. They reported that the denitrification rate is influenced by the amount and composition of organic matter sinking to the sediments. The greater the supply of organic matter, the faster the rate of denitrification.

There is a significant relationship between NO_3^- reduction rate and temperature in the sediments at Lauffen (Fig. 4.5). NO_3^- reduction rate reached a maximum of 1.1 mmol/m^2 d at 20 °C and was only 0.69 mmol/m^2 d at 10 °C (Table 4.1). In a laboratory experiment, Sagemann et al. (1994) found that the denitrification rate of

NO_3^- in the sediments from the Weser Estuary is related to temperature. High temperatures cause not only a high diffusion rate of NO_3^-, but also high bacteria activity, consequently high mineralization of organic matter (Holdren and Armstrong 1986).

In summary, the denitrification rate appears to be dependent on the amount of organic matter reaching the sediments, the availability of NO_3^-, and temperature.

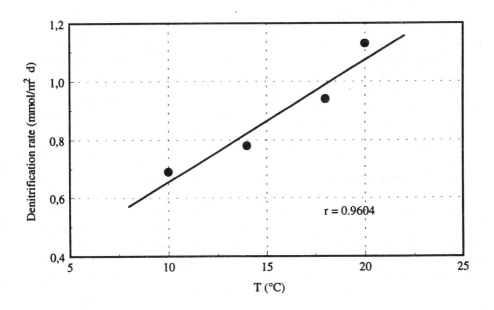

Fig. 4.5. Relationship between denitrification rate and temperature

4.1.3 Cycling of manganese in the sediments

At all sites dissolved Mn^{2+} in the overlying water was below the detection limit (1 µM). At sites Lauffen and Kochendorf, porewater Mn^{2+} could not be detected in upper 0-6 cm. Below this layer, Mn^{2+} concentrations increased rapidly to 25 µM at 16 cm depth. At site Wieblingen, the increase of Mn^{2+} began directly below the sediment-water interface, the maximum value was 62 µM at 10 cm depth (Fig. 4.6).

Similar porewater profiles were found in the tributaries of the Neckar River (Fig. 4.7). It is noticeable that the depth, where Mn^{2+} was released into the porewater, was identical with the beginning of NO_3^- reduction. This indicates that NO_3^- reduction and Mn oxide reduction occurred simultaneously in the sediments.

In the Neckar River sediments average particulate Mn was highest at Wieblingen (650 mg/kg), followed by Kochendorf (560 mg/kg) and at Lauffen (410 mg/kg). A slight accumulation of particulate Mn at the surface sediment layer was identified at Lauffen and Wieblingen. At Kochendorf, particulate Mn concentrations ranged from 360 to 760 mg/kg through the sediment core (Fig. 4.6).

The porewater and solid phase profiles of Mn are similar to those reported from other sediments (Gendron et al. 1986; Johnson et al. 1992; Matsunaga et al. 1993). These profiles demonstrate transport processes of Mn during the mineralization of organic matter. Under anoxic conditions, Mn oxides are reduced and Mn^{2+} is released into porewater:

$$(CH_2O)_{113}(NH_3)_{15}(H_3PO_4) + 248.5\ MnO_2 + 497\ H^+$$
$$\rightarrow 248.5\ Mn^{2+} + 113\ CO_2 + 7.5\ N_2 + H_3PO_4 + 384\ H_2O$$

The release of Mn^{2+} is limited by either the quantity of organic matter or abundance of available oxides. The high concentrations of dissolved Mn^{2+} in the Neckar River sediments at Wieblingen and in the Enz River sediments are probably related to high contents of Mn in the solid phase (Fig. 4.6c, 4.7b).

Dissolved Mn^{2+} in the porewater diffuses towards the surface sediment layer due to concentration gradients. As Mn^{2+} is reoxidized as Mn oxides and immobilized in the surface oxic layer, an accumulation of particulate Mn in this layer should be expected.

$$Mn^{2+} + O_2 \rightarrow MnO_2\downarrow$$

The diagenetic enrichment of Mn in the surface layer has been reported from marine sediments (Aller 1980b; Pedersen and Price 1982; Dahmke et al. 1991). However, only a slight accumulation of Mn was found in the sediments at Lauffen, Wieblingen, and in the Enz River sediments, but no peak of Mn was measured at Kochendorf and in the Elsenz River sediments. This might be the result of sediment mixing processes by bioturbation and hydrodynamic forces.

The concentrations of Mn^{2+} in the porewater are influenced not only by redox reactions, but also by a series of precipitation and dissolution of Mn minerals (Matsunaga et al. 1993):

rhodochrosite: $\quad Mn^{2+} + CO_3^{2-} \rightarrow MnCO_3 \downarrow$

alabandite : $\quad Mn^{2+} + HS^- \rightarrow MnS \downarrow + H^+$

The Saturation Index (SI) of possible minerals was calculated by ion activity products (IAP) in comparison to thermodynamic equilibrium constant. Corrections were made for ionic strength. For instance, the SI of rhodochrosite ($MnCO_3$) is defined as:

$$SI = Log\ (IAP\ /K_{sp})$$
$$= Log(a_{Mn}\ a_{CO3}\ /\ K_{sp})$$
$$= Log\ \{([Mn^{2+}]\gamma^{2+}\ [CO_3^{2-}]\gamma^{2-})\ /\ K_{sp}\}$$

where IAP: Ion Activity Products,

K_{sp}: equilibrium constant for $MnCO_3$,

a_{Mn}, a_{CO3}: activity of Mn^{2+} and CO_3^{2-} in solution,

γ^{2+}, γ^{2-}: activity coefficient used to correct the concentration data to activity,

$[Mn^{2+}], [CO_3^{2-}]$: the measured concentration of Mn^{2+} and CO_3^{2-}.

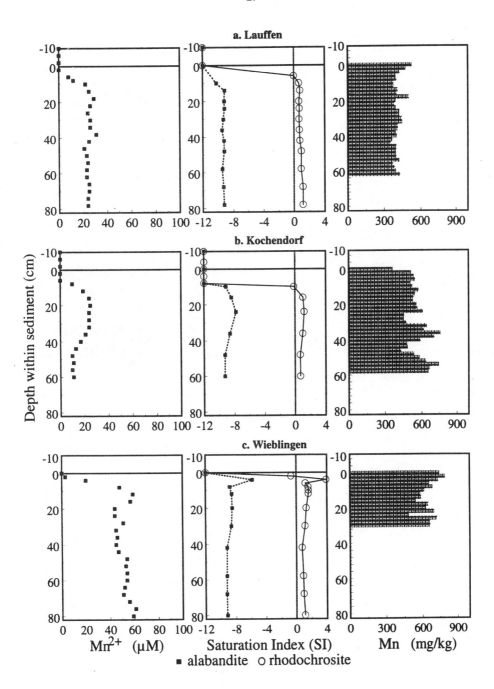

Fig. 4.6. Dissolved Mn^{2+}, SI of Mn minerals, and Mn in the sediments of the Neckar River

28

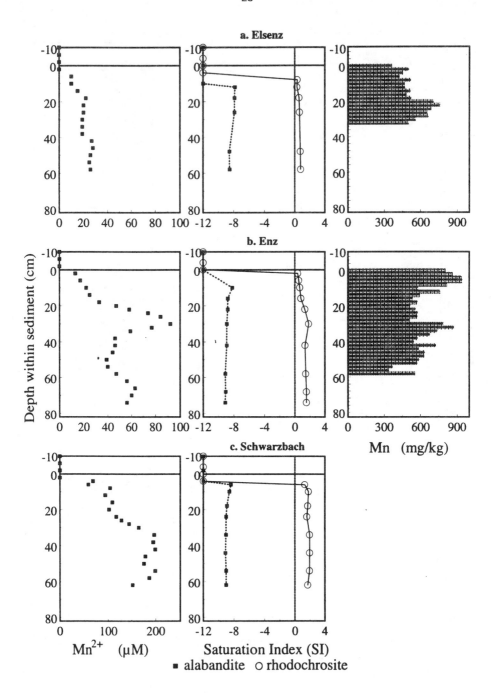

Fig. 4.7. Dissolved Mn^{2+}, SI of Mn minerals, and Mn in the sediments of the tributaries

The computer program MINTEQA2 was used to solve a set of simultaneous non-linear equations. Parameters for these equations are set by the thermodynamic equilibrium and mass balance. The principal advantage of this model is to consider the effect of complexation by ligands and competition of other minerals.

The saturation calculation indicates that the porewater is slightly supersaturated (SI ≈ 2) in respect to rhodochrosite in the sediments of the study area (Fig. 4.6, 4.7). Using X-ray diffraction analysis for Baltic Sea sediment samples, Suess (1979) identified that the dominant crystalline phase was a mixed Mn-carbonate: $(Mn_{0.85}Ca_{10}Mg_{0.05})CO_3$, rather than pure $MnCO_3$. This is in agreement with the results from Aller (1980b), Pedersen and Price (1982). As no thermodynamic parameters are available for the mixed Mn-carbonate, solid solution of other components such as Ca and Mg considerably complicates the calculation of saturation. It is reasonable to assume ideal behavior (pure mineral) for the component which has a mole fraction $N > 0.9$ (Pedersen and Price 1982). The rhodochrosite and/or Mn-Ca-carbonate appear to control the dissolved Mn^{2+} in the anoxic sediments. Suess (1979) found that Mn-carbonate in the Baltic Sea sediments primarily occurs within thin layers consisting of a porous, spongy framework of amorphous silica. It seems that precipitation of the manganous carbonate is possibly enhanced by the presence of grain surfaces which encourage seed mineral formation or nucleation. The porewater was strongly undersaturated in respect to alabandite (MnS; Fig. 4.6 and 4.7), so control by alabandite is not possible.

4.1.4 Cycling of iron in the sediments

At Lauffen, dissolved Fe^{2+} increased until 0.20 mM between 10 and 14 cm depth. Below this layer, it varied between 0.16 and 0.30 mM. At Kochendorf, Fe^{2+} increased slowly between 8 and 50 cm depth. Likewise, an increase of Fe^{2+} was also found in the sediments at Wieblingen.

Low Fe^{2+} concentrations were measured in the Elsenz River sediments. Dissolved Fe^{2+} increased with depth to 0.03 mM at 48 cm depth (Fig. 4.9a). The maximum value of Fe^{2+} was 0.53 mM at 32 cm depth in the Enz River sediments (Fig. 4.9b). In the Schwarzbach River sediments, Fe^{2+} concentrations increased continuously to 0.90 mM at 42 cm depth. Below this depth, Fe^{2+} decreased with depth (Fig. 4.9c).

No significant difference of particulate Fe with depth was found at all sites. The average values were the same at Lauffen, Kochendorf, and at Wieblingen (3.0 %; Fig. 4.8). Higher contents of particulate Fe were measured in the sediments of the tributaries: 4.5 % in the Elsenz River and 3.4 % in the Enz River.

Similar to Mn cycling, the behavior of Fe is controlled by redox reactions. Ferric Fe is present as oxyhydroxides in the oxic waters. It is reduced to Fe^{2+} in the anoxic sediments. Unlike other electron acceptors such as O_2, NO_3^-, and SO_4^{2-}, Mn and Fe oxides remain high concentrations through anoxic sediments. This is probably due to their extremely low solubility, which may make them less biologically available than other oxidants.

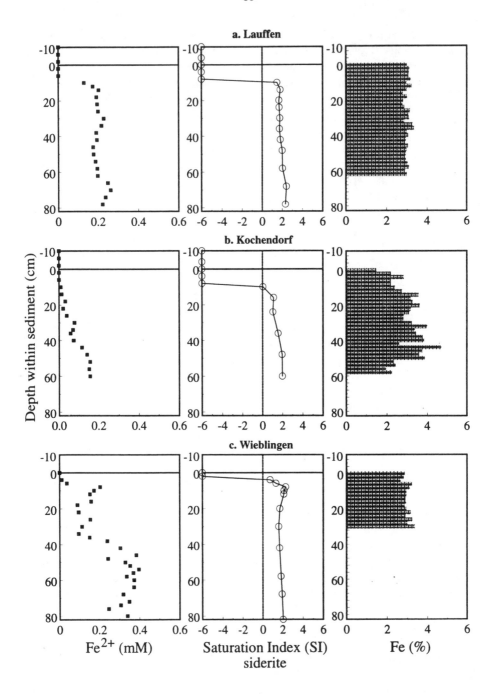

Fig. 4.8. Dissolved Fe^{2+}, SI of siderite, and Fe in the sediments of the Neckar River

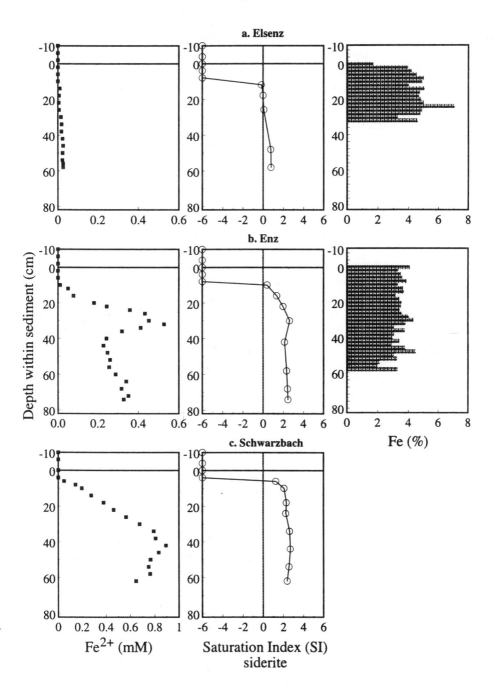

Fig. 4.9. Dissolved Fe^{2+}, SI of siderite, and Fe in the sediments of the tributaries

Because of its concentration gradient, Fe^{2+} in the sediments diffused upward, and was reoxidized as Fe oxide. This process can be described as follows:

$$(CH_2O)_{113}(NH_3)_{15}(H_3PO_4) + 452\ FeOOH + 904\ H^+$$
$$\rightarrow 452\ Fe^{2+} + 113\ CO_2 + 15\ NH_3 + H_3PO_4 + 791\ H_2O$$
$$4\ Fe^{2+} + O_2 + 10\ H_2O \rightarrow 4\ Fe(OH)_3 + 8\ H^+$$

However, a significant accumulation of Fe at the surface sediments was not found at all sites. This can be attributed to the geogenic abundance of Fe in the sediments. Assuming all of Fe^{2+} diffusing through the porewater was reoxidized as Fe oxides at the surface sediments, the flux of Fe^{2+} can be estimated by Fick's First Law:

$$F = \phi^2\ D_{fe}(\partial C / \partial x),$$

where D_{fe} : diffusion coefficient of Fe^{2+} in water,
$$D_{fe} = 0.503\ cm^2 / d.\ at\ 18\ °C \quad (Li\ and\ Gregory\ 1974),$$

ϕ: porosity,

$\partial C / \partial x$: porewater Fe^{2+} concentration gradient at depth x.

The maximum calculated flux of Fe^{2+} is 0.47 mg/cm^2 y in the sediments at Wieblingen. Assuming a sedimentation rate of 1 cm/y, the increase in particulate Fe due to the oxidation of Fe^{2+} is estimated to be roughly 0.02%. This value is very small relative to the particulate Fe contents (3.0%) in the sediments. Therefore, it is not possible to demonstrate the effect of the early diagenetic processes from the distribution of particulate Fe in the sediments. It seems that the sedimentation of Fe oxides, rather than a diagenetic reaction controls the concentrations of particulate Fe. In addition, the mixing of the sediments resulting from bioturbation and/or hydrodynamic forces may have an important effect in the distribution of particulate Fe.

Like Mn^{2+}, porewater Fe^{2+} in anoxic sediments is also influenced by precipitation and dissolution of Fe minerals:

siderite: $\qquad\qquad\qquad\quad Fe^{2+} + CO_3^{2-} \rightarrow FeCO_3$

FeS nH$_2$O(amorphous) $\qquad Fe^{2+} + HS^- + nH_2O \rightarrow FeS\ nH_2O + H^+$

makinawite $\qquad\qquad\qquad Fe^{2+} + HS^- \rightarrow FeS + H^+$

vivianite $\qquad\quad 3\ Fe^{2+} + 2\ PO_4^{3-} + 8\ H_2O \rightarrow Fe_3(PO_4)_2\ 8H_2O$

ΣH_2S concentrations in the sediments of the study area were lower than 1 μM. According to Berner (1981), the Neckar sediments can be described as methanic environment. In such sediments the coexistence of siderite, vivianite, and iron-sulfide should be expected. Siderite might precipitate and control dissolved Fe^{2+} in the absence of abundant sulfide.

The existence of siderite is supported by x-ray diffraction determinations in the sediments of Lac Leman, Switzerland (Nembrini et al. 1982), and in the Baltic Sea sediments (Suess 1979). The porewater in the sediments of our study area is slightly supersaturated with respect to siderite (Fig. 4.8 and 4.9). Such a supersaturation seems to be a common feature and has been reported in several other studies of anoxic sediments (Emerson 1976; Postma 1981). As the precipitation of siderite is an extremely slow process, the formation of siderite appears to occur by precipitation only from strongly supersaturated porewater (Postma 1981).in addition, the complexation of Fe with organic colloids (e.g. humic substances) may lead to a supersaturation of siderite (Emerson 1976; Aller 1980b; Elderfield 1981), which unfortunately is not considered in the calculation.

4.1.5 Sulfate reduction

SO_4^{2-} profiles in porewater show a depletion in SO_4^{2-} below the sediment-water interface. In the Neckar River, SO_4^{2-} concentrations in the overlying water varied from 0.79 to 1.26 mM. They decreased rapidly in the sediments at depths between 2 and 40 cm (Fig. 4.10a). Below 40 cm, the SO_4^{2-} concentrations were less than 0.05 mM.

The SO_4^{2-} concentrations in the overlying water of the tributaries were lower than those in the Neckar River. In the Enz River water 0.50-0.65 mM of SO_4^{2-} were measured. A decrease of SO_4^{2-} in the porewater began at 22 cm depth (Fig. 4.10b). In the Enz River, the SO_4^{2-} concentrations decreased rapidly from 0.43 mM in the bottom water to about 0.01 mM at 20 cm depth. In the sediments of the Schwarzbach River, SO_4^{2-} was consumed between 4 and 20 cm depth. Below this zone, the SO_4^{2-} concentrations were lower than 0.01 mM. SO_4^{2-} profiles in the sediments at Lauffen show no significant seasonal variation (Fig. 4.10c).

Numerous studies have attested that the availability of SO_4^{2-} and organic matter in sediments control SO_4^{2-} reduction (Mountfort and Asher 1981; Billen 1982; Jørgensen and Sørensen 1985). Freshwater has usually low concentrations of SO_4^{2-} (0.1-0.2 mM) compared to sea water (20-30 mM, Capone and Kiene 1988). A large difference in SO_4^{2-} concentrations between freshwater and marine sediments results in a different speciation of sulfur in the sediments. Generally, SO_4^{2-} reduction leads to HS^- production in the sediments:

$$(CH_2O)_{113}(NH_3)_{15}H_3PO_4 + 56.5 \; SO_4^{2-}$$
$$\rightarrow 113 \; HCO_3^- + 15 \; NH_3 + H_3PO_4 + 56.5 \; HS^- + 56.5 \; H^+$$

HS^- may react with Fe^{2+}, yielding Fe monosulfides and pyrite (FeS_2). Therefore, strong SO_4^{2-} reduction accounts for the accumulation of inorganic reduced sulfur in marine sediments. Since freshwater has low SO_4^{2-} concentrations (0.2 mM, Capone and Kiene 1988) and a high input of organic matter, organic sulfur, instead of inorganic sulfur, is the main fraction of sulfur in the sediments. As a result of human activities (e.g. agricultural runoff, industrial and communal discharge), the concentrations of SO_4^{2-} in the Neckar River and its tributaries were 0.40-1.26 mM. The strong SO_4^{2-} reduction in the sediments of the study area is clearly reflected by the sharp decrease in SO_4^{2-} concentrations. In addition, the black color of the sediment cores is an indicator of amorphous $FeS \cdot nH_2O$ and makinawite (FeS). As heavy metals can be precipitated with HS^- to form highly insoluble metal-sulfides or be

34

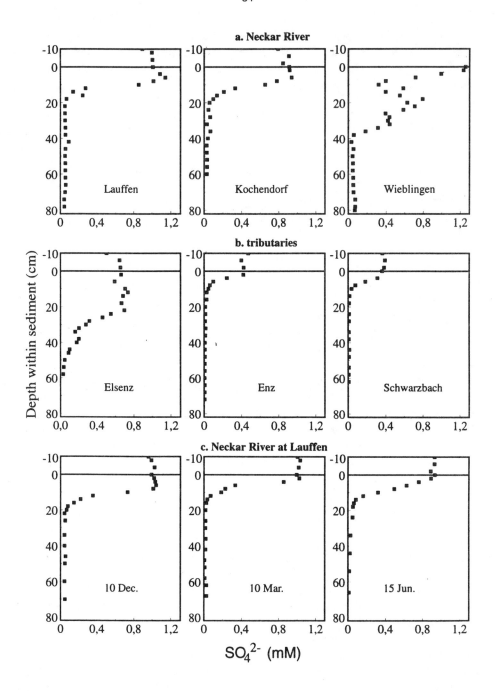

Fig. 4.10. Sulfate profiles in the porewater from the Neckar River and its tributaries

4.1.6 Ammonium production

In the overlying water of the Neckar River, NH_4^+ concentrations ranged from 0.01 to 0.04 mM, and increased rapidly with depth in the porewater. The maximum values of NH_4^+ in the sediments were 15 mM at Lauffen, 3.6 mM at Kochendorf, and 8.4 mM at Wieblingen (Fig. 4.11a).

Only a slight increase of NH_4^+ was found in the sediments of the Elsenz River. The maximum value was 0.44 mM. In the Enz River sediments, a rapid increase of NH_4^+ was measured below 20 cm depth. NH_4^+ concentrations in the sediments of the Schwarzbach River show a different tendency: they increased to 50 cm depth, and subsequently decreased with depth (Fig. 4.11b). No distinct seasonal changes were apparent from the porewater NH_4^+ profiles obtained in the sediments at Lauffen (Fig. 4.11c).

In anoxic sediments, SO_4^{2-} reduction and CH_4 fermentation are the predominant processes in the mineralization of organic matter (Aller 1980a; Billen 1982; Klump and Martens 1987; Barbanti et al. 1992a):

$$(CH_2O)_{113}(NH_3)_{15}H_3PO_4 + 56.5\ SO_4^{2-}$$
$$\rightarrow 113\ HCO_3^- + 15\ NH_3 + H_3PO_4 + 56.5\ HS^- + 56.5\ H^+$$
$$(CH_2O)_{113}(NH_3)_{15}H_3PO_4 \rightarrow 56.5\ CH_4 + 56.5\ CO_2 + 15\ NH_3 + H_3PO_4$$

The NH_4^+ profiles from different sites vary considerably. In most cases, particularly in the sediments at Wieblingen, the NH_4^+ profiles can be separated into two zones: in the SO_4^{2-} reduction zone and in the zone below the SO_4^{2-} reduction zone (Fig. 4.11). The SO_4^{2-} reduction zone is clearly defined by exponential decrease in SO_4^{2-} concentrations. Below this zone, there is no further contribution from SO_4^{2-} reduction, yet NH_4^+ concentrations continued to increase more or less rapidly, suggesting contribution from CH_4 fermentation. In the sediments of Lake Washington, Kuivila at al. (1989) found that CH_4 production rate was low until SO_4^{2-} concentrations decreased below 0.03 mM. Only then did CH_4 production start to increase. The spatial separation of SO_4^{2-} reduction and CH_4 fermentation indicates competition of both processes in anoxic freshwater sediments. In a laboratory experiment, Winfrey and Zeikus (1977) reported that when SO_4^{2-} was added to lake sediment samples, CH_4 fermentation was suppressed until SO_4^{2-} was consumed. In another experiment, when molybdate - a specific inhibitor of SO_4^{2-} reduction - was added to the sediments, methane fermentation increased under the presence of SO_4^{2-} (Capone et al. 1983). This can be interpreted as a result of competition between methane producing- and SO_4^{2-}-reducing bacteria for acetate and hydrogen. Kuivila et al. (1989) reported, for the first time, the separation of methane production and SO_4^{2-} reduction in low-SO_4^{2-} (0.1 mM) freshwater sediments of Lake Washington. Likewise, the separate NH_4^+ profiles in the porewater suggest that SO_4^{2-} reduction and CH_4 fermentation are spatially separated in the sediments of the study area, which have high SO_4^{2-} concentrations. This situation is observed only in highly eutrophic and polluted sediments where organicmatter input to the sediments

36

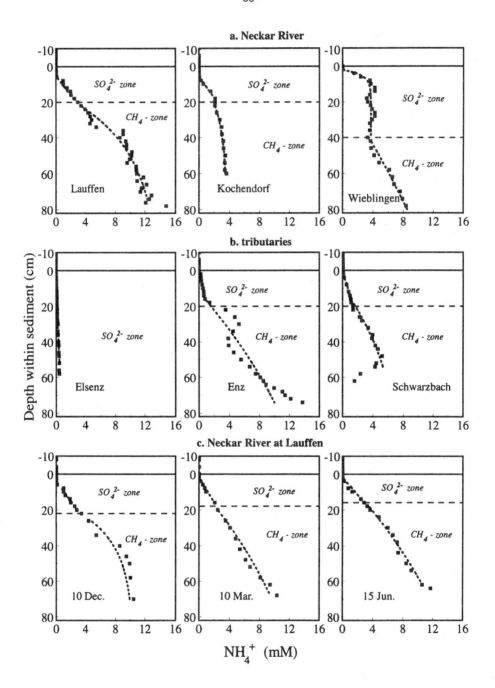

Fig. 4.11. Ammonium profiles in the porewater from the Neckar River and its tributaries

is large. Despite a high rate of SO_4^{2-} reduction, there is still sufficient labile organic matter present after SO_4^{2-} depletion to support CH_4 fermentation. In the Elsenz River, where SO_4^{2-} reduction process occurred throughout the sediment core, only slight increase of NH_4^+ was measured in the sediments, which was probably resulted from SO_4^{2-} reduction.

The decomposition rate of organic matter depends on the availability of labile organic matter in the sediments. Since the availability of labile organic matter decreases with depth, an increase of the decomposition rate might be observed only in the uppermost sediment layer. This may explain the similar NH_4^+ profile from different seasons at site Lauffen.

4.1.7 Alkalinity, Ca, Mg, and conductivity in the porewater

In the sediments of the Neckar River, alkalinity increased rapidly with depth at sites Lauffen and Wieblingen. The maximum values were 44 meq/l (at 78 cm) at Lauffen, and 33 meq/l (at 80 cm) at Wieblingen. At Kochendorf, the increase of alkalinity was slower and the maximum value was 16 meq/l at 60 cm depth (Fig. 4.12a).

In the Elsenz River, only a slight increase of alkalinity was identified in the sediments. The maximum value (8.4 meq/l) was measured at 58 cm depth. In the Enz River, the maximum value (24 meq/l) was found at the deepest sampling depth (Fig. 4.12b). No significant seasonal variation of alkalinity profiles was found in the sediments at Lauffen (Fig. 4.12c). Like NH_4^+ production, alkalinity production can be separated into two zones: in the SO_4^{2-} reduction zone and in the CH_4 fermentation zone, indicating the different alkalinity generation processes in the sediments.

The porewater profiles of Ca^{2+} and Mg^{2+} in the sediments are illustrated in Fig. 4.13 and 4.14. An increase with depth can be observed at all sites. The gradient for both parameters in the porewater was smallest in the Elsenz River sediments, and slightly higher in the sediments of the Enz River and Schwarzbach River, whereas the highest was measured in the Neckar River at Lauffen and Wieblingen. This pattern coincides with the behavior of NH_4^+ concentrations and alkalinity. In contrast, carbonate contents in the sediments of the study area have no clear tendency with depth. In the sediments at Kochendorf, 25 % of the carbonate were measured. The carbonate content was 6.5 % in the Elsenz River sediments and 6.8 % in the Enz River sediments, respectively. The carbonate contents were lower compared to those in the Neckar River (Fig. 4.15).

38

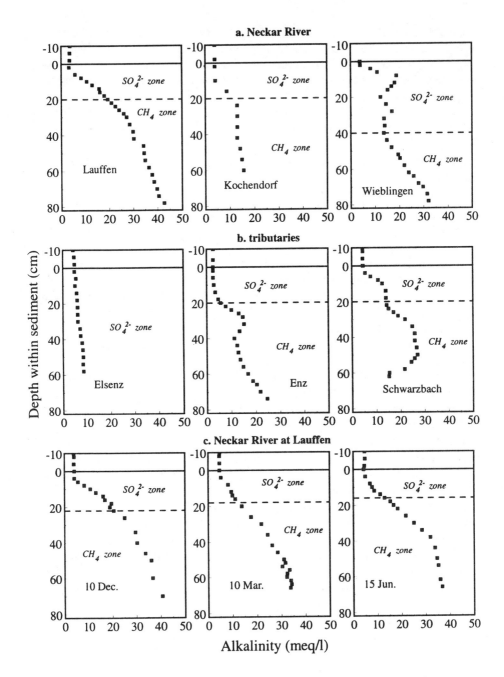

Fig. 4.12. Alkalinity profiles in the porewater from the Neckar River and its tributaries

39

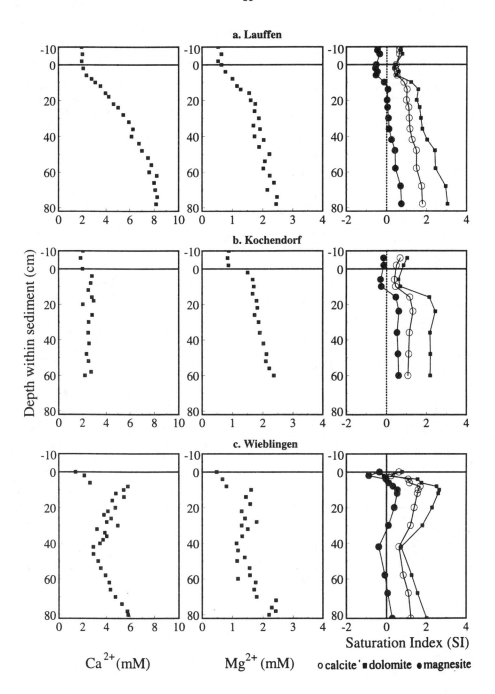

Fig. 4.13. Dissolved Ca^{2+}, Mg^{2+}, and SI of their minerals in the porewater of the Neckar River

40

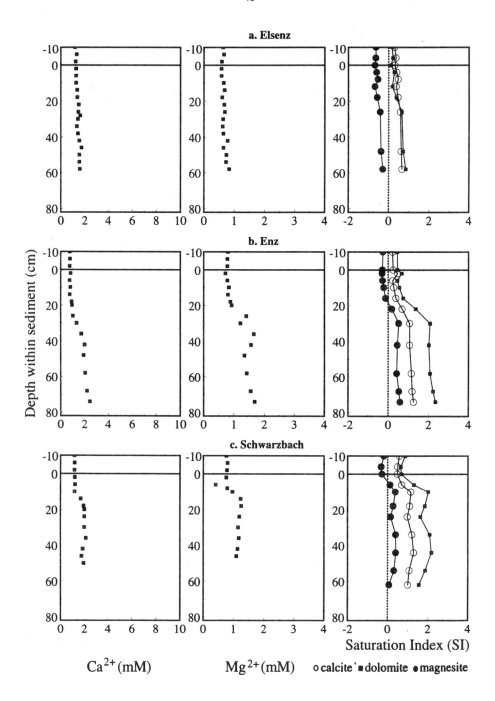

Fig. 4.14. Dissolved Ca^{2+}, Mg^{2+}, and SI of their minerals in the porewater of the tributaries

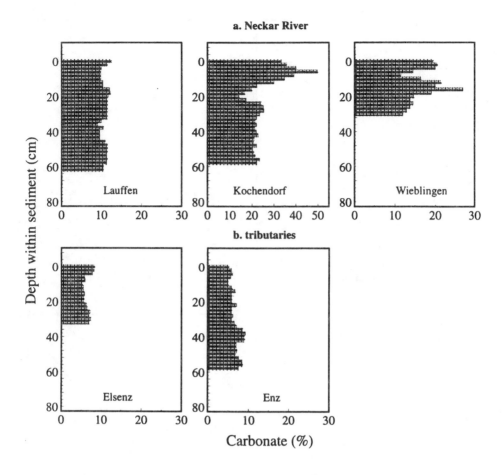

Fig. 4.15. Carbonate concentrations in the sediments of the Neckar River and its tributaries

The mineralization of organic matter releases CO_2 and NH_3 into porewater of anoxic sediments. This process can change the chemical equilibrium (e.q. precipitation/dissolution, adsorption/desorption of Ca and Mg), and consequently control the distribution of these elements between sediments and water:

$$CaCO_3 + CO_2 + H_2O \rightarrow CaHCO_3^+ + HCO_3^-$$
$$MgCO_3 + CO_2 + H_2O \rightarrow MgHCO_3^+ + HCO_3^-$$
$$NH_3 + CO_2 + H_2O \rightarrow NH_4^+ + HCO_3^-$$
$$Me^+\text{-clay} + NH_3 + CO_2 + H_2O \rightarrow NH_4^+\text{-clay} + Me^+ + HCO_3^-$$

where Me^+: 0.5 Ca^{2+}, 0.5 Mg^{2+}.

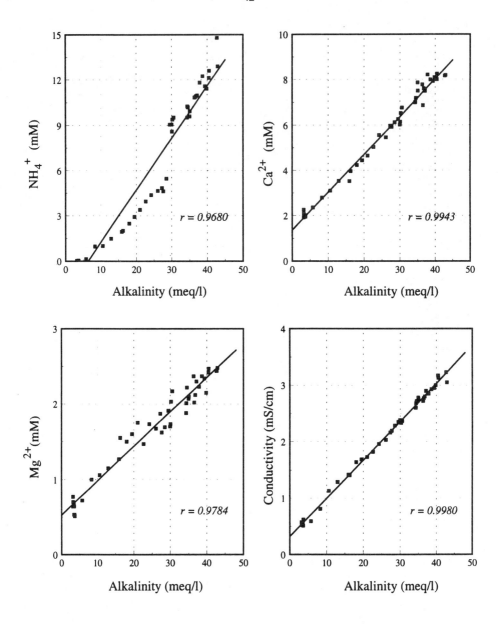

Fig. 4.16. Relationship between NH_4^+, Ca^{2+}, Mg^{2+}, conductivity, and alkalinity in the porewater at Lauffen

Carbonate dissolution, ion exchanges with Ca^{2+} and Mg^{2+}, and NH_4^+ production during the mineralization of organic matter contribute to the alkalinity in the sediments. The significant correlations between alkalinity, NH_4^+, Ca^{2+}, and Mg^{2+} in the porewater at Lauffen confirm this mechanism (Fig. 4.16). Calcite ($CaCO_3$) may precipitate in the bottom water and deposit in the sediments (Sigg 1986; Sigg et al. 1987).

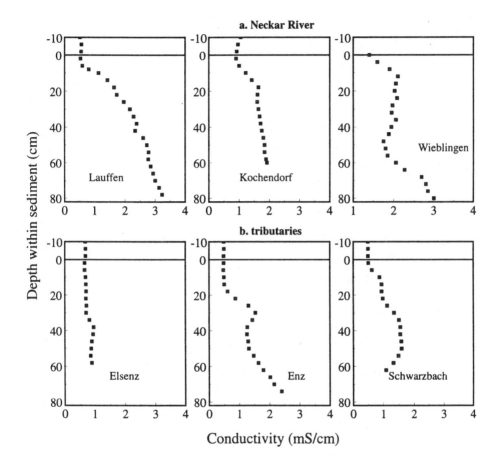

Fig. 4.17. Conductivity profiles in the porewater of the Neckar River and its tributaries

Since the Ca:Mg ratio in the bottom water is < 2, a formation of dolomite and magnesite in the bottom water cannot be expected (Müller et al. 1987). Instead, low-Mg calcite might be the predominant precipitate of carbonate. The increase of dissolved Ca^{2+} and Mg^{2+} in the porewater coincides with the increase of alkalinity. This demonstrates that low-Mg calcite in the sediments dissolved through the increase of CO_2, which is released by the mineralization of organic matter. In the sediments of the study area, porewater is supersaturated with respect to calcite

Matisoff et al. (1981), which is maintained by organic and colloidal complexes. In addition, nucleation may play an important role in the precipitation of calcite.

Conductivity of porewater is an index of ion activity (sum of ion concentrations). At all sites, conductivity increased correspondingly with depth, reflecting its association with the mineralization of organic matter (Fig. 4.17)

4.1.8 Phosphorus in the sediments and porewater

In the sediments at Lauffen, after a sharp increase of PO_4^{3-} concentrations from 0.003 mM in the bottom water to 0.23 mM at 8 cm depth, the concentrations of PO_4^{3-} decreased to about 0.015-0.056 mM. The peak of PO_4^{3-} concentration (0.16 mM) lay in 10 cm depth at Wieblingen. At Kochendorf, the maximum PO_4^{3-} concentration (0.11 mM) was found in 30 cm depth (Fig. 4.18).

In the Elsenz River sediments, PO_4^{3-} concentration increased slowly to 26 cm depth, and varied between 0.094 and 0.13 mM below this depth. No clear decrease was found in the sediments. In the Enz River sediments, PO_4^{3-} increased to a maximum value (0.21 mM) at 24 cm depth and then decreased to a relatively constant value (0.10 mM). Similar profiles were found in the Schwarzbach river sediments (Fig. 4.19).

The contents of particulate P varied between 0.21 % and 0.38 % in the sediments at Lauffen, and 0.12 %-0.51 % at Wieblingen. Only 0.10-0.19 % of P was measured in the sediments at Kochendorf. In the Enz River sediments, P content decreased from 0.32 % at surface sediment layer to 0.15 % at 32 cm depth. Below this depth, it ranged from 0.17 % to 0.29 % (Fig. 3.18, 3.19).

In natural waters, sedimentation of biologically produced P, biogenic coprecipitation of P with $CaCO_3$, and adsorption of P by Fe and Mn oxides account for the dominant transfer of PO_4^{3-} from bottom water to sediments. After deposition of these particles, PO_4^{3-} is released not only by the decomposition of organic matter, but also by the reduction of Fe and Mn oxides (Froelich et al. 1979; Pettersson 1986; Dahmke et al. 1991; Sundby et al. 1992). Further evidence of PO_4^{3-} release with the reduction of Fe and Mn oxides was reported by Krom and Berner (1981). In their experiment, PO_4^{3-} was released into the solution during the reduction of ferric iron by H_2S. They found that below the Fe oxide reduction zone, inorganic particulate P remained constant, while organic P continued to decrease, suggesting a biogenic source of dissolved PO_4^{3-} in the porewater.

PO_4^{3-} and Fe^{2+} diffuse towards the surface sediments following the concentration gradients. At the surface oxic layer, PO_4^{3-} is readsorbed by freshly formed Fe and Mn oxides. On the otherhand, the fraction of organic phosphorus decreased with depth following the decomposition of organic matter. Therefore, P-rich sediment layers should be expected at the surface sediments, as reported by Krom and Berner (1981) for Long Island Sound sediments, by Balzer (1989) for the Kiel Bight, and by Sundby et al. (1992) for the Gulf of St. Lawrence. However, such a P-rich surface

45

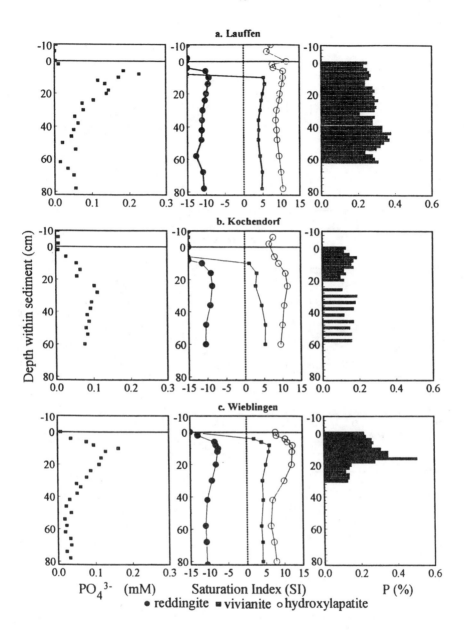

Fig. 4.18. Phosphate, SI of P minerals, and P in the sediments of the Neckar River

46

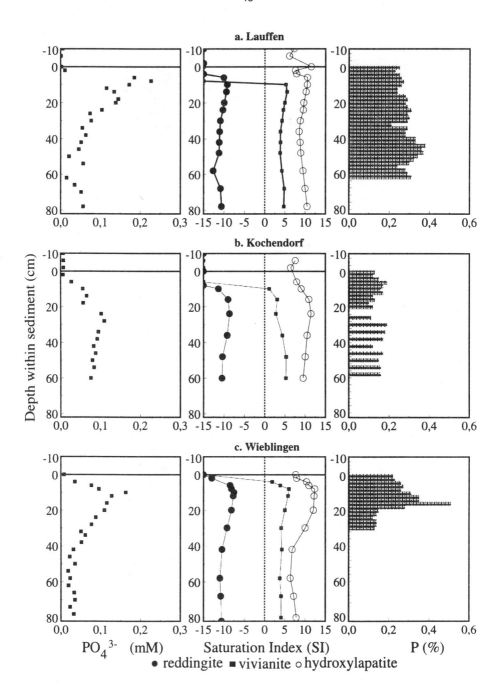

Fig. 4.19. Phosphate, SI of P minerals, and P in the sediments of the tributaries

layer is not found in the Neckar River sediments, the resuspension and mixing of the sediments by bioturbation and hydrodynamic forces (e.g. flood) may be responsible for this feature.

Phosphorus minerals such as vivianite [$Fe_3(PO_4)_2$ $8H_2O$] and reddingite [$Mn_3(PO_4)_2$ $3H_2O$] may control PO_4^{3-} concentrations in anoxic sediments (Nriagu and Dell 1974; Emerson 1976; Aller 1980a; Nembrini et al. 1982; Holdren and Armstrong 1986):

$$\text{vivianite:} \quad 3\,Fe^{2+} + 2\,PO_4^{3-} + 8\,H_2O \rightarrow Fe_3(PO_4)_2\, 8H_2O$$

$$\text{hydroxylapatite:} \quad 5\,Ca^{2+} + 3\,PO_4^{3-} + OH^- \rightarrow Ca_5(PO_4)_3(OH)$$

$$\text{reddingite:} \quad 3\,Mn^{2+} + 2\,PO_4^{3-} + 3\,H_2O \rightarrow Mn_3(PO_4)_2\, 3H_2O$$

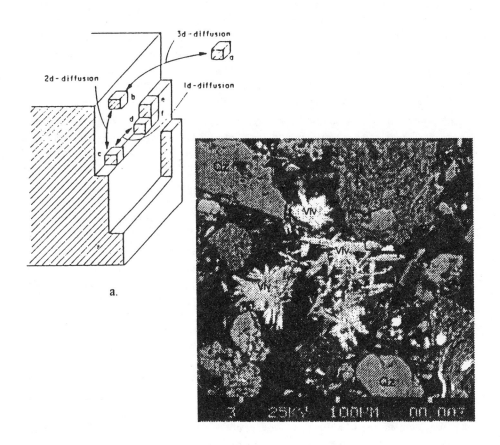

a.   b.

Fig. 4.20a. The various steps in the precipitation processes

b. BSEM-Photomicrograph-Vivianite crystals from Neckar River sediments
(Viv: vivianite, Qz: quartz)

Generally, four steps can be distinguished in the attachment of solutes to a solid surface (Fig. 4.20a):

a-b) diffusion in the solution boundary layer adjacent to the surface (bulk diffusion).

b) adsorption reaction with the solid surface.

b-c) migration on the surface to a step edge (surface diffusion).

c-d) migration along a step edge to a kink (edge diffusion).

Any of these steps, alone or in combination, may limit the kinetics of precipitation. When step 1 only is limiting, the kinetics is said to be transport controlled; when steps 2, 3, and/or 4 only are limiting, the kinetics is said to be controlled by a surface reaction.

In the study area, a strong supersaturation of hydroxylapatite (SI > 6) is found at all sites in the bottom water and porewater. As described by Holdren and Armstrong (1986), formation of hydroxylapatite is often kinetically inhibited and has no influence on the behavior of PO_4^{3-}, unless large quantities of relatively fine grained calcite are present. In the River Bog sediments, Postma (1981) has not found any evidence of Ca-rich phosphate. These results demonstrate that the precipitation of hydroxylapatite is controlled by surface reactions rather than by a transport process.

The porewater is supersaturated with respect to vivianite (SI $\approx$ 6) below the depth where Fe^{2+} was released into porewater (Fig. 4.18 and 4.19). Vivianite has been reported as the stable iron phosphate phase in a number of lake sediments, where the porewater is often supersaturated with vivianite (Nriagu and Dell 1974; Holdren and Armstrong 1986). This may suggest the slow precipitation kinetics of this mineral. The existence of vivianite in the Enz River sediments was confirmed by Peula (Fig. 4.20b, unpublished data) using microscope analysis. Vivianite crystals were also identified by X-ray diffraction for the River Bog sediments in Denmark (Postma 1981). According to Morel and Hering (1993), the precipitation rate is limited by the rate of surface diffusion and edge diffusion controlled by the amounts of 'step' and 'kink' sites (Fig. 4.20). Usually, these sites can propagate in a spiral during crystal growth at high supersaturation. The precipitation rate is then directly proportional to the supersaturation. This explanation is consistent with the observations that the formation of vivianite is apparently precipitated from strong supersaturated solutions (Nriagu and Dell 1974; Emerson 1976; Nembrini et al. 1982; Holdren and Armstrong 1986).

The possibility of reddingite [$Mn_3(PO_4)_23H_2O$] formation has been suggested for the sediments from Lake Erie (Matisoff et al. 1980). The porewater in the sediments of the Neckar River and its tributaries is strong undersaturated with reddingite (SI < - 7), suggesting that reddingite formation is not a significant factor controlling PO_4^{3-} concentration in the sediments (Fig 4.18 and 4.19).

In general, the porewater PO_4^{3-} concentrations illustrate the rapid release of PO_4^{3-} into the porewater and the slow precipitation of P minerals. The PO_4^{3-} peaks below the sediment-water interface can be explained as a results of PO_4^{3-} release from organic and inorganic 2bound P. The decrease of the PO_4^{3-} concentrations in the deeper sediments could result from the formation of phosphorus minerals such as vivianite.

4.1.9 Effect of oxygen-depleting substances (NH_4^+, Fe^{2+})

The oxic surface layer acts as a trap for PO_4^{3-} and NH_4^+ released by the mineralization of organic matter. When this oxic layer is eliminated because of O_2-depletion in the overlying column, or by strong resuspension of the sediments due to high diacharge or dredging, the buffer capacity of the sediments will be destroyed, and a release of PO_4^{3-} and NH_4^+ into the overlying water can be expected. In extreme situations this can lead to an ecological catastrophe for the aquatic ecosystem.

In April 1992, an extensive fish kill occurred in the Schwarzbacher River at Neckarbischofsheim, directly near a discharge conduit of a sewage treatment plant. To find the reason for this ecological disturbance, two "peepers" were installed in this area. Site A was located at upstream from the discharge pipe of the treatment plant, while site B was at downstream from the pipe.

The measured porewater profiles at site A and site B are shown in Fig. 4.21. At site B, NO_3^- and SO_4^{2-} decreased more rapidly than at site A. Dissolved Mn^{2+} and Fe^{2+} concentrations were about two times higher than at site A. As a product of the decomposition of organic matter, the highest NH_4^+ concentration (5.2 mM) was found at site B, which was about four times higher than at site A. The porewater profiles measured at site A and B show different decomposition processes of organic matter. Since the water flows slowly at site B, more suspended particles deposit at site B compared to site A. Therefore, higher input of labile organic matter to the surface sediments of site B may result in higher decomposition rates. This explanation is supported by higher concentrations of NH_4^+ and alkalinity in the sediments at site B compared to site A.

The mineralization of organic matter in anoxic sediments results in the formation of NH_4^+, Fe^{2+}, and sulfide. Generally, these oxygen-depleting substances diffuse upward and are oxidized in the oxic surface layer. They have no serious effect on the ecosystems due to their low concentrations. If the oxic layer is destroyed (e.g. resuspension of sediments by flood or dredging), the surface sediments may lose their scavenger function and release NH_4^+, Fe^{2+}, and sulfides into the overlying water. As a result, the oxygen demand in the overlying water would increase:

$$NH_4^+ + 2\,O_2 \rightarrow NO_3^- + H_2O + 2\,H^+$$
$$4\,Fe^{2+} + O_2 + 10\,H_2O \rightarrow 4\,Fe(OH)_3 + 8\,H^+$$
$$4\,FeS + 9\,O_2 + 10\,H_2O \rightarrow 4\,Fe(OH)_3 + 4\,H_2SO_4$$

1 g of NH_4^+ can consume 4.5 g of O_2. According to Imhoff (1976), 1 liter river sludge can consume within 24 hours an amount of O_2 equal to the quantity of dissolved O_2 in 100 liter of overlying water. Müller and Schleichert (1977) measured the amounts of suspended sediment and dissolved O_2 in the Rhine, where a fish kill occurred between 6-8 June, 1971. As a result of heavy rains in the catchment areas of several tributaries, the water level rose unusually fast

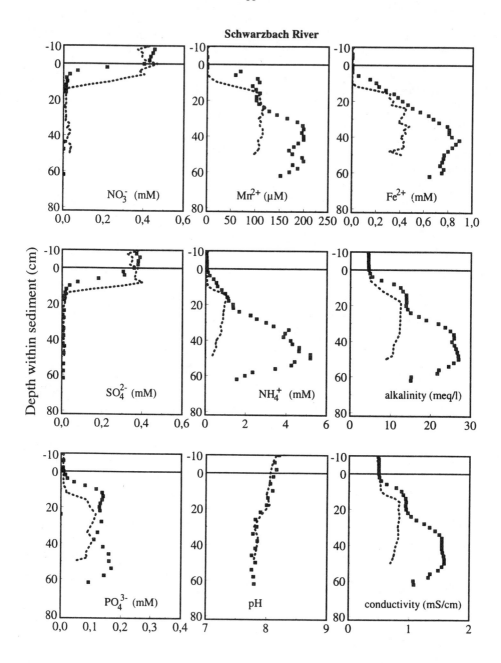

Fig. 4.21. Porewater profiles at station A (dashed line) and B (■) in the Schwarzbach River

before the fish kill in the Schwarzbach River, a lock was opened for repairs. This resulted in a strong resuspension of the sediment and may have caused a release of large amounts of O_2-depletion substances (NH_4^+, Fe^{2+}). Furthermore, NH_4^+ alone is a toxic substance for the nerve system of fish. It can lead to muscle cramp, and consequently fish kill. (Baur 1980).

Normally, the increased O_2 demand and high NH_4^+ concentrations during flood is compensated by dilution through the large water runoff, which has no significant effect on aquatic organisms in the bottom water. It should be noted that both fish kills (in the Rhine and the Schwarzbach River) occurred in June and April, respectively. The fish kills did not occur with every flood. In summer, the water level is lower than in winter, and dissolved O_2 concentrations are lower due to biological O_2 demand. As a result, the buffer capacity of the river for O_2 depletion is lower in summer, especially in a small river such as the Schwarzbach River. Therefore, a resuspension of the sediments by human activities (e.g. dredging) should be avoided in summer. The best management for such system would be to stabilize the sediment structure and to minimize the fluctuation in oxidation-reduction conditions in the sediments.

4.1.10 Sinks or sources of sediments for bromine

At all sites, Br^- in the overlying water is below the detection limit (0.25 µM). The concentrations of Br^- in the porewater showed a tendency to increase with depth. The gradient of Br^- was smallest in the sediments of the Elsenz River (Maximal value: 0.63 µM), slightly higher in the Schwarzbach River sediments, whereas in the sediments of the Neckar River exhibited a comparingly high gradient (Fig. 4.22). The maximal value was 9.26 µM and 9.01µM at Lauffen and Wieblingen, respectively. In the Enz River sediments, Br^- increased from undetected in the bottom water to a maximal value of 7.76 µM at 74 cm depth.

Particulate Br contents in the Neckar River sediments at Lauffen varied between 3.9 and 4.6 mg/kg, which is in agreement with data from freshwater lacustrine sediments containing Br in the range of 6-18 mg/kg (Mun and Bazilevich 1962).

The Br^- profiles were very similar to the profiles of NH_4^+ and alkalinity in the same site. It is well known that the mineralization of organic matter leads to the production of alkalinity and NH_4^+, their concentrations are then a measure of the strength and/or duration of the degradation process, and of the composition and concentration of organic matter available in the sediments. The very high positive correlation between Br^-, NH_4^+, and alkalinity leads to the conclusion that bromine, originally a constituent of the organic matter in the sediments, is released as Br^- during early diagenetic processes (Fig. 4.23).

Only few analyses exist for Br^- concentrations in nonmarine plants. Coastal waters are enriched with $CHBr_3$ and CH_2Br_2 due in part to their production by marine macroalgae and possibly by marine microbes (Manley et al. 1992). Some organobromine compounds such as $CHBr_3$, CH_2Br_2, and $CHBrCl_2$ were measured in marine macroalgae tissues (Gschwend et al. 1985) The function of these compounds

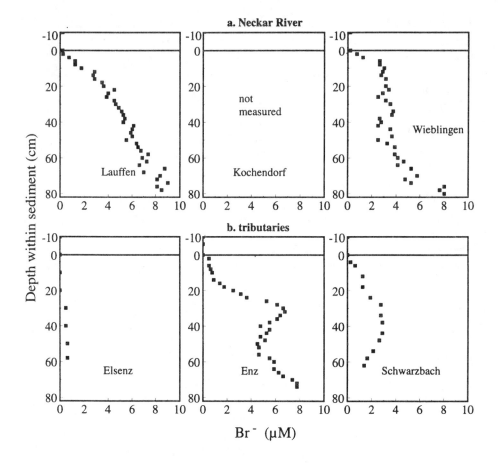

Fig. 4.22. Bromide in the porewaters of the Neckar River and its tributaries

compounds in the algae is unknown. Since many volatile halogenatal organic compounds exhibit antibiotic properties, one can speculate that these chemicals may be involved in epiphytic control.

No data at all could be found for freshwater phytoplankton, which is believed to make up an essential portion of the organic matter in the sediments. For grass and clover, 17-119 and 19-52 mg/kg were reported, respectively (Kabata-Pendias and Pendias 1984). However, these values should be considered with utmost care, since no clear differentiation between water-soluble and -insoluble Br has been made.

Gschwend et al. (1985) believed that there is no known major mechanisms for the removal of organobromine from sea water other than exchange with atmosphere. The result in the study area reveals that sediments act both as sinks and sources for Br. Br⁻

released from the organic matter will finally reach the bottom water by diffusion and/or compaction of the sediments. The Sediments therefore play an important role in the cycling of this element in the biogeosphere.

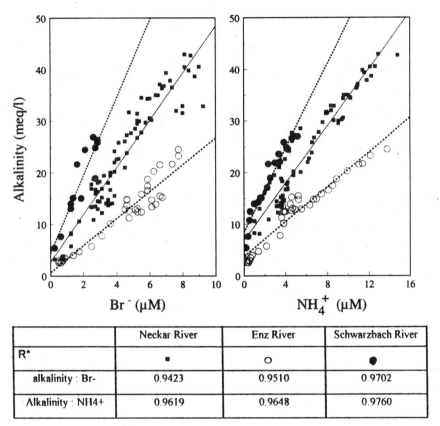

	Neckar River	Enz River	Schwarzbach River
R*	■	O	●
alkalinity : Br-	0.9423	0.9510	0.9702
Alkalinity : NH4+	0.9619	0.9648	0.9760

* R: correlations coefficient

Fig. 4.23. Relationship between bromide, ammonium, and alkalinity.

4.2 Distribution of heavy metals between sediments and porewater

4.2.1 Copper

Considering the concentrations of particulate Cu in the sediments of the Neckar River, two zones can be separated: a relatively less polluted layer and a heavily polluted layer (Table 4.2). The interface between the two zones lay at 20 cm depth at Lauffen and 22 cm depth at Kochendorf (Fig. 4.24). At Wieblingen, the younger, less polluted sediment layer was located between 0 and 6 cm due to a removal of sediment by high water discharge. High concentrations of particulate Cu were measured in the Enz River sediments below 22 cm depth. The average value in the Elsenz River sediments was 87 mg/kg, No significant difference with depth was found through the sediment core (Fig. 4.25).

Table 4.2. Distribution of Cu in the sediments (< 20 μm) of the Neckar River and its tributaries

	Depth cm	Average mg/kg	Min. mg/kg	Max. mg/kg
Lauffen	0-20	96	89	110
	20-64	320	200	430
Kochendorf	0-22	56	46	61
	22-60	140	100	170
Wieblingen	0-6	78	74	79
	6-32	150	72	230
Elsenz	0-34	87	48	120
Enz	0-22	170	150	200
	22-60	360	160	560

The high concentrations of Cu in the sediments did not coincide with a release of dissolved Cu into the porewater. Maximum values of dissolved Cu (53-210 nM) were measured at the sediment-water interface in the Neckar River. A sharp decrease of the dissolved Cu was identified below the interface. In the older sediment layer with higher concentrations in particulate Cu, dissolved Cu was less than or equal to 5 nM (Fig. 4.24).

A remarkable peak of dissolved Cu (350 nM) was found at the sediment-water interface in the Enz River. In the Elsenz River and the Schwarzbach River, the dissolved Cu concentrations were higher in the overlying water than in the porewater (Fig.4.25). The rapid decrease of Cu in the porewater with depth occurred below the sediment-water interface.

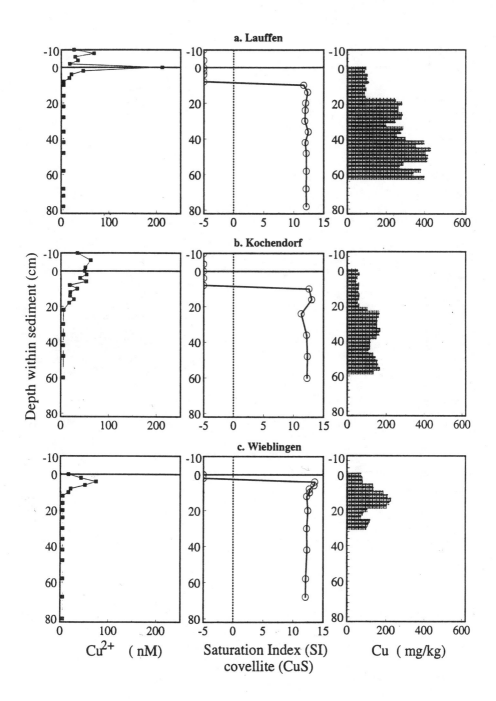

Fig. 4.24. Dissolved Cu, SI of CuS, and Cu in the sediments (< 20 µm) of the Neckar River

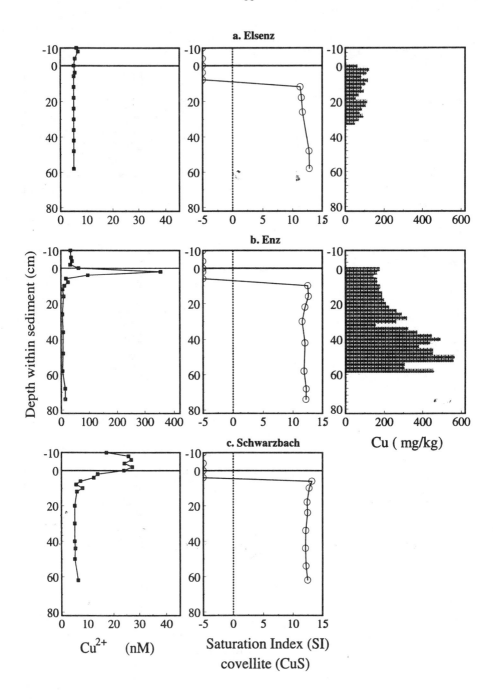

Fig. 4.25. Dissolved Cu, SI of CuS, and Cu in the sediments (< 20 μm) of the tributaries

Dissolved Cu in the bottom water and in the porewater was lower than the international standard for drinking water (16 μM, WHO). Therefore, under present physicochemical conditions, no direct influence of dissolved Cu on the quality of ground water and bottom water is to be expected.

SO_4^{2-} reduction during the mineralization of organic matter leads to the production of HS^-. Highly insoluble metal-sulfides can form when traces of the HS^- ion are present in the porewater. Consequently, the mobility of these metals can be significantly lowered (Förstner and Wittmann 1979; Carignan and Nriagu 1985; Dahmke et al. 1991; Williams 1992; Morse and Arakaki 1993). Another possibility is the coprecipitation of Cu with FeS:

$$Cu^{2+} + HS^- \rightarrow CuS + H^+$$
$$Fe^{2+} + HS^- \rightarrow FeS + H^+$$

With electron microprobe analysis, Lee and Kittrick (1984a) found that sulfur (33%) and Fe (27 %) are the most important associations for particulate Cu in contaminated soil samples. X-ray dot pictures showed that Cu and the associated elements are scattered rather than concentrated along the edge of the particles. This indicates that Cu has been precipitated with sulfur as CuS or coprecipitated as chalcopyrite ($CuFeS_2$), instead of being adsorbed on the particle surface. In a laboratory experiment, Wallmann (1992a) found that Cu concentration in a suspension solution began to decrease when HS^- was detected in the solution, suggesting the formation of CuS in the solution.

HS^- produced by the reduction of SO_4^{2-} rapidly precipitates Fe^{2+} in the form of FeS nH_2O (amorphous) or makinawite (FeS), which are further converted into pyrite. Carignan and Lean (1991) reported that H_2S concentrations in Williams Bay sediments are controlled by the formation of iron monosulfides. They found that the Ion Activity Products (IAP) of Fe^{2+} and S^{2-} remains just above the solubility product (Ksp) of mackinawite (FeS). Thus, Fe^{2+} concentrations in the porewater can be used to estimate HS^- concentrations in anoxic sediments (Billen 1982, Morfett et al. 1988). The formation of iron monosulfides is confirmed by the black color of the sediment cores in the study area. The calculated HS^- values range between 10^{-7} and 10^{-8} M, which is in good agreement with the data from literature (Fernex et al. 1986; Carignan and Lean 1991).

The Saturation Index (SI) for CuS (covellite) is presented in Fig.4.24 and 4.25. Strong supersaturation was found at all sites. The mostly probable explanation is that Cu can be strongly complexed by organic colloids, which was not considered in the calculation. This explanation is consistent with the observation that Cu has a strong affinity for organic ligands. Boulegue et al. (1982) found that 90 % of dissolved Cu was complexed as Cu-organo-sulfur complex in the porewater of Great Marsh, Delaware. Similar results were reported by Elderfield (1981) for the porewater of Narragansett Bay sediments.

Douglas et al. (1986) measured 22-67 % of dissolved Cu in the porewater from Narragansett Bay. He found that the dissolved organic carbon (DOC) increased with depth due to the mineralization of organic matter. This could enhance the mobility of Cu by organic-Cu complexation. Indeed, dissolved Cu decreased with depth. It appears that the interaction of Cu with DOC is not sufficiently strong to prevent Cu-sulfide formation. The decrease of dissolved Cu in the porewater of the study area indicates that Cu was fixed in the particulate phase. From the porewater profiles, a release of Cu from the sediments into the overlying water can not be expected. In contrast, the sediments appear to act as a sink rather than a source for Cu.

4.2.2 Cadmium

Like Cu profiles, a sharp increase of particulate Cd concentrations was identified at 20 cm depth at Lauffen and at 24 cm depth at Kochendorf. Because of the removal of the sediments by flood, the less polluted layer lay at a depth of 0-6 cm at Wieblingen (Table 4.3; Fig. 4.26).

In the Elsenz River sediments, particulate Cd varied between 0.51 and 2.9 mg/kg. No significant difference of the Cd concentrations with depth was found in the sediments. In the Enz River sediments, particulate Cd concentrations increased from 4.7 mg/kg at 22 cm depth to 46 mg/kg at 54 cm depth.

Table 4.3. Distribution of Cd in the porewater and sediments (<20 µm) of the Neckar River and its tributaries.

	Depth	Average mg/kg	Min. mg/kg	Max. mg/kg	Dissolved Cd nM *
Lauffen	0-20	1.6	1.2	2.4	-
	20-64	49	25	84	-
Kochendorf	0-22	1.6	1.2	2.2	0.89 (2 cm)
					0.62 (6 cm)
					0.45 (10 cm)
	22-60	3.7	1.4	7.3	
Wieblingen	0-6	2.8	1.7	4.5	0.62 (6-8 cm)
	6-32	15	1.9	34	
Elsenz	0-34	1.2	0.51	2.9	
Enz	0-22	3.0	1.5	6.0	
	22-60	25	4.7	46	

*Cd could only be detected in some porewater samples, values in brackets indicate the respective depth.

The very high concentrations of Cd in the older sediment layer are consistent with the industrial activities in the Neckar River drainage area before 1973. Because of the large amounts of Cd discharge into the Neckar River and its toxicity, Cd contamination in the sediments and in the aquatic organisms was extensively reported from this area (Förstner and Müller 1973; Müller and Prosi 1977, 1978; Müller 1980,

1981, 1986; Müller et al. 1993). Inspite of a significant decrease of the heavy metal pollution, the sediments are still moderately to strongly polluted in respect to Cd in the younger sediment layer. New studies of the sediment assessment show that the concentrations of particulate Cd remained constant or increased slightly compared to 1985 (Müller et al. 1993). This can be attributed to the resuspension and transport of the old sediments by hydrodynamic forces such as flood, which frequently occurs in the drainage area of the Neckar River.

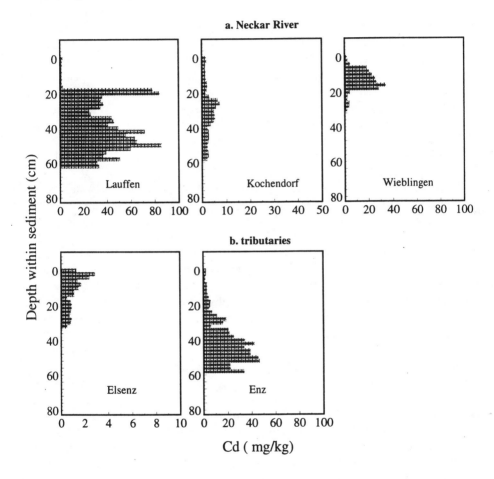

Fig. 4.26. Cd in the sediments (< 20 µm) of the Neckar River and its tributaries

In contrast to the high contents of Cd in solid phase, dissolved Cd could only be detected in some samples in the uppermost layer of the Neckar River sediments (Table.4.3). The concentrations of the dissolved Cd were lower than international standard for drinking water (44 nM, WHO). The dissolved Cd appears to be rapidly removed from solution to particles. Furthermore, this removal occurred just below the sediment-water interface. Comparing the Saturation Index of CdS ($4.2 < SI < 4.8$) and the distribution of Cd between sediments and porewater, it seems that the depletion of dissolved Cd in the anoxic sediments results from the precipitation of CdS or coprecipitation with FeS:

$$Cd^{2+} + HS^- \rightarrow CdS\downarrow + H^+$$

Evidence of CdS formation in anoxic sediments was found by determination of electron beam microprobe analysis in a heavy metal polluted harbor sediments from Michigan City (Lee and Kittrick 1984b). They reported that 89 % of the particulate Cd is associated with sulfur in the sample. X-ray dot pictures showed that Cd, like Cu, is dispersed rather than concentrated along the edge of particles. It seems that Cd has been coprecipitated with sulfur, instead of being adsorbed. In the sediments of the Northwest Mediterranean Continental Shelf, Fernex et al. (1986) found the existence of polymetallic sulfides, with which Cd could be associated. The formation of CdS was also confirmed by laboratory experiments. In a suspension experiment of the sediment sample from the Elbe River, Wallmann (1992a) found that the concentrations of dissolved Cd decreased when HS^- was detected in the solution. Thus, the precipitation of CdS is suggested.

However, some studies lead to different conclusions. In the porewater of the Continental Margin of Central California, McCorkle and Klinkhammer (1990) showed that dissolved Cd concentrations increased from 0.6 to 1.5 nM in the top 0.5 cm of the sediments, and then rapidly decreased to values of approximately 0.12 nM. Because there was no significant SO_4^{2-} reduction and no HS^- was detected in the sediments, the decrease of dissolved Cd in the sediments can be explained by scavenging onto freshly formed iron oxide surfaces. Tessier et al. (1985) proposed that trace metal concentrations are controlled by adsorption onto solid substances such as Fe oxides above the SO_4^{2-} reduction zone. In these cases, the low availability of HS^- in the sediments may limit the formation of metal-sulfides. The sharp decreases of SO_4^{2-} in the Neckar River (from 1.26 mM to 0.05 mM) and its tributaries (from 0.66 mM to 0.01 mM) show a strong SO_4^{2-} reduction and HS^- production within the sediments. This may lead to the precipitation of CdS and coprecipitation of Cd with FeS. As a result, the high concentrations of Cd in the sediments do not result in a release of Cd into the porewater. This interpretation is in agreement with other observations (Carignan and Nriagu 1985; Carignan and Tessier 1985; Morfett et al. 1988; Dahmke et al. 1991; Williams 1992). Therefore, the formation of metal sulfide may act as a sink for Cd in the heavily contaminated sediments of the study area.

4.2.3 Lead

Particulate Pb concentrations in the sediments of the Neckar River are shown in Table 4.4. High concentrations of Pb in the older sediment layer are related to those of Cu and Cd, indicating the influence of human activities (Fig. 4.27).

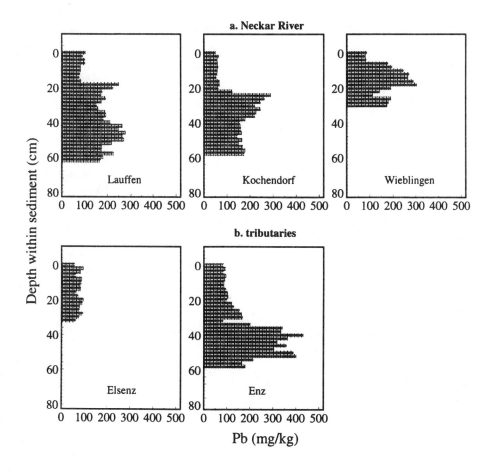

Fig. 4.27. Pb in the sediments (< 20 μm) of the Neckar River and its tributaries

In most porewater samples dissolved Pb was below the detection limit (2.5 nM). In the samples near the sediment-water interface, Pb concentrations varied from 2.5 nM to 6.1 nM, distinctively lower than international standard for drinking water (240 nM).

The solubility calculations indicate that PbS in these samples is supersaturated (3.6 < SI < 4.8):

$$Pb^{2+} + HS^- \rightarrow PbS \downarrow + H^+$$

The precipitation of PbS and the coprecipitation of Pb with FeS are also supported by the strong reduction of SO_4^{2-} in the sediments and the depletion of dissolved Pb in the porewater.

Table 4.4. Distribution of Pb in the porewater and sediments (< 20 μm) of the Neckar River and its tributaries

	Depth cm	Average mg/kg	Min. mg/kg	Max. mg/kg	Dissolved Pb nM*
Lauffen	0-20	90	78	100	-
	20-64	210	150	280	-
Kochendorf	0-22	60	51	67	2.8 (0 cm)
					3.8 (2 cm)
					4.3 (6 cm)
					2.5 (10 cm)
					2.9 (10 cm)
	22-60	190	120	290	-
Wieblingen	0-6	84	80	87	2.5 (0 cm)
					6.1 (4 cm)
					5.0 (6 cm)
					3.5 (8 cm)
	6-32	210	110	300	-
Elsenz	0-34	80	57	95	-
Enz	0-22	93	86	100	-
	22-60	250	84	430	-

*Pb could only be detected in some porewater samples, values in brackets indicate the respective depth.

4.2.4 Zinc

Depth profiles of particulate Zn demonstrate a significant decrease of Zn discharge since 1973. Exception for the Elsenz River sediments, particulate Zn concentrations in the older sediment layer doubled those of the younger sediment layer. No significant difference of particulate Zn was found in the Elsenz River sediments (Table. 4.5).

Notable peaks of dissolved Zn at the sediment-water interface were measured in the Neckar River (Fig. 4.28). A sharp decrease of dissolved Zn concentrations was found between 0 and 20 cm depth. The highest concentration of dissolved Zn (3.1×10^3 nM) was measured at the sediment-water interface in the Enz River. The concentrations of Zn decreased dramatically to 130 nM at 6 cm depth and remained at a fairly low level (60-100 nM). Low concentrations of dissolved Zn were measured in the bottom water of the Schwarzbach River and Elsenz River. Within the sediments they also decreased with depth (Fig. 4.29).

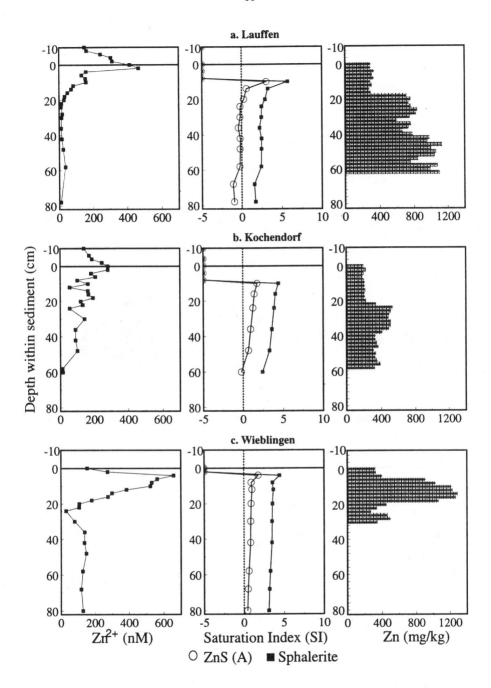

Fig. 4.28. Dissolved Zn, SI of Zn minerals, and Zn in the sediments (< 20 μm) of the Neckar

64

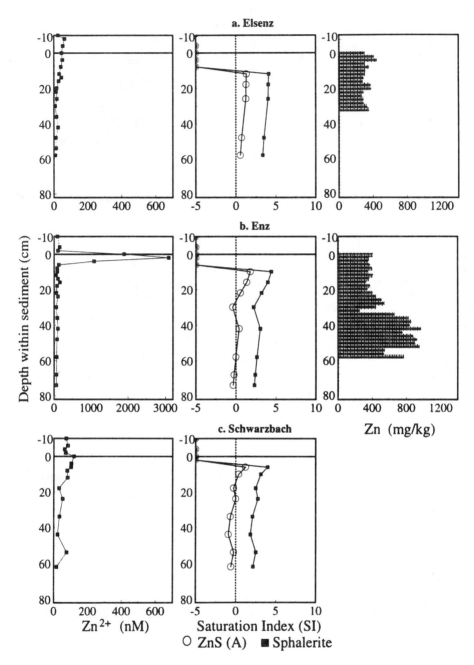

Fig. 4.29. Dissolved Zn, SI of Zn minerals, and Zn in the sediments (< 20 μm) of the tributaries

Table 4.5. Distribution of Zn in the sediments (< 20 μm) of the Neckar River and its tributaries

	Depth cm	Average mg/kg	Min. mg/kg	Max. mg/kg
Lauffen	0-20	300	270	330
	20-64	860	600	1100
Kochendorf	0-22	200	180	220
	22-60	410	220	540
Wieblingen	0-6	350	320	400
	6-32	800	270	1300
Elsenz	0-34	320	280	440
Enz	0-22	370	330	410
	22-60	670	250	970

Such a decrease of dissolved Zn was reported by Carignan and Nriagu (1985) for acid lakes in Canada, and by Dahmke et al. (1991) for the Weser estuary, Germany. The most probable explanation is the formation of ZnS:

$$Zn^{2+} + HS^- \rightarrow ZnS\downarrow + H^+$$

Using an electron microprobe, Lee and Kittrick (1984a) found that 65-94 % of particulate Zn in an anoxic sediment sample from the Michigan City harbor is associated with sulfur. Furthermore, Zn and sulfur are distributed rather than concentrated on the edge of particles. This indicates that Zn has been coprecipitated as ZnS rather than adsorbed in the sediments. This interpretation is also confirmed by the results in the sediments of the study area. First, the saturation calculation indicates that the porewater is slightly supersaturated in respect to sphalerite (ZnS) and in equilibrium with ZnS (amorphous). The supersaturation can be explained by slow precipitation kinetics and/or formation of organic complexes. van den Berg and Dharmvanij (1984) reported that the organic fraction of Zn accounts for 93-98 % of the dissolved Zn in the porewater from the Mersey River estuary in England. Secondly, the decrease of dissolved Zn in the sediments of the study area indicates that the sediments act as a sink rather than a source. It is apparent that ZnS precipitation controls the distribution of Zn between sediments and porewater.

4.2.5 Chromium

The concentrations of particulate Cr are shown in Fig. 4.30 and 4.31. Cr profiles show different concentrations between the younger sediment layer and the older sediment layer (Table. 4.6), indicating an anthropogenic source.

Dissolved Cr concentrations in the surface waters varied from 8 nM to 12 nM and decreased below the detection limit (6 nM) in the sediment layer between 0-20 cm. Below this layer, Cr concentrations increased with depth (Fig. 4.30 and 4.31). Like

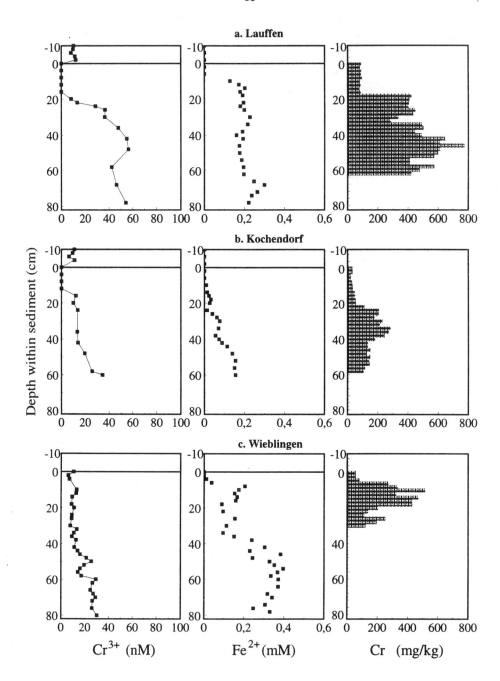

Fig. 4.30. Porewater Cr and Fe, and Cr in the sediments (< 20 μm) of the Neckar River

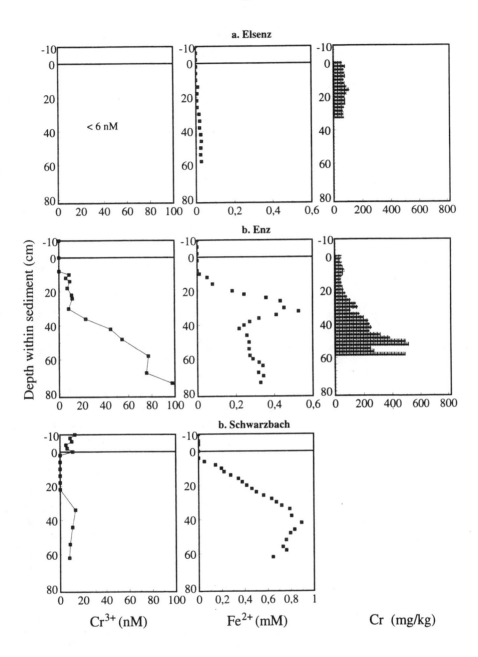

Fig. 4.31. Porewater Cr and Fe, and Cr in the sediments (< 20 μm) of the tributaries

other heavy metals, the concentrations of dissolved Cr in the bottom water and in the porewater were lower than the international standard for drinking water (960 nM, WHO). This reflects that porewater Cr has no direct influence on the quality of the overlying water and groundwater.

Table 4.6. Distribution of Cr in the sediments (< 20 µm) of the Neckar River and its tributaries

	Depth cm	Average mg/kg	Min. mg/kg	Max. mg/kg
Lauffen	0-20	85	76	95
	20-64	490	280	780
Kochendorf	0-22	36	20	51
	22-60	180	52	280
Wieblingen	0-6	63	54	80
	6-32	290	110	520
Elsenz	0-34	78	60	110
Enz	0-22	51	39	66
	22-60	240	60	520

Cr is a redox-sensitive element and is found as Cr(VI) or Cr(III) in natural waters, depending on the redox conditions. Balistrieri et al.(1992) found that Cr exists as CrO_4^{2-} in oxic water of the Lake Sammamish, Washington. In the bottom water of the study area, Cr is stable as CrO_4^{2-}. The CrO_4^{2-} diffused into the uppermost sediment layer is probably reduced to Cr(III).

As Cr(III) can strongly bind to particles (Fe/Mn oxides, clay minerals), dissolved Cr could not be detected in this layer. Below this surface layer, Fe and Mn oxides are reduced due to the mineralization of organic matter, a release of Cr into the porewater can be expected. The shapes of porewater Cr profile are similar to the profiles of dissolved Fe. In the Elsenz River sediments, no dissolved Cr was detected. Only 0.1-0.3 mM of dissolved Fe was measured in the porewater, which was 5-10 fold lower than that at other site. In this regard, the cycling of Cr seems to be more closely linked to the cycling of Fe rather than that of Mn. This mechanism is supported by the results from Johnson et al. (1992).

Whether or not sulfide was present, Salomons et al. (1987) found no difference in Cr concentrations in a laboratory adsorption experiment. In this case, dissolved Cr could be controlled by adsorption processes. Johnson et al. (1992) reported that the reduced Cr is not in true solution as Cr(III), but is present as colloidal Cr(III), probably due to the strong tendency of Cr(III) to organic colloids. Since the organic colloids generally increase with depth through the decomposition of organic matter (Vuynovich 1989), the increase of dissolved Cr in the sediments can also be explained as a complex of Cr-organic colloids. Douglas et al. (1986) reported that the total and organic Cr in the porewater of Narragansett Bay sediments increase with depth, suggesting the formation of organic-Cr-complexes.

4.2.6 Cobalt

In contrast to other heavy metals, the concentrations of particulate Co in the sediments of the study area display no difference between the younger and the older sediment layer (Table 4.7). The sediments are relatively unpolluted in respect to Co (Fig. 4.32 and 4.33).

Co in the surface water was below the detection limit (20 nM) at all sites. Porewater Co increased with depth below the sediment-water interface, which coincides with the increase of Fe^{2+} and Mn^{2+} in the porewater.

The strong affinity of Co for Fe/Mn oxides has been well described in the scientific literature (Gendron et al. 1986; Shaw et al. 1990; Williams 1992; Wallmann 1992b). Balistrieri et al. (1992) reported that temporal changes of dissolved Co in Lake Sammamish, Washington are very similar to those of dissolved Mn. Based on a laboratory experiment, Wallmann (1990) demonstrated that dissolved Co is released with the reduction of Fe/Mn oxides. Co in the solution decreased with the increase of HS^- concentrations, suggesting the formation of CoS. This is supported by the saturation index of CoS in the solution:

$$Co^{2+} + HS^- \rightarrow CoS\downarrow + H^+$$

The saturation calculation for CoS indicates that CoS was in equilibrium with the porewater (Fig. 4.32 and 4.33) at all sites. It appears that Co cycling in the sediments is related to Fe and Mn oxide reduction and CoS precipitation. In the bottom water, Co is most likely associated with Fe/Mn oxides and is removed from the water by settling particles. After the deposition, dissolved Co is released during the reduction of Fe and Mn oxides and is controlled by the precipitation of CoS.

Table 4.7. Distribution of Co in the sediments (< 20 μm) of the Neckar River and its tributaries

	Depth cm	Average mg/kg	Min. mg/kg	Max. mg/kg
Lauffen	0-20	12	10	14
	20-64	13	9.4	16
Kochendorf	0-22	7.8	5.8	12
	22-60	9.5	8.7	10
Wieblingen	0-6	8.0	8.0	8.0
	6-32	9.3	8.0	10
Elsenz	0-34	12	8.3	13
Enz	0-22	10	9.3	11
	22-60	9.1	6.2	11

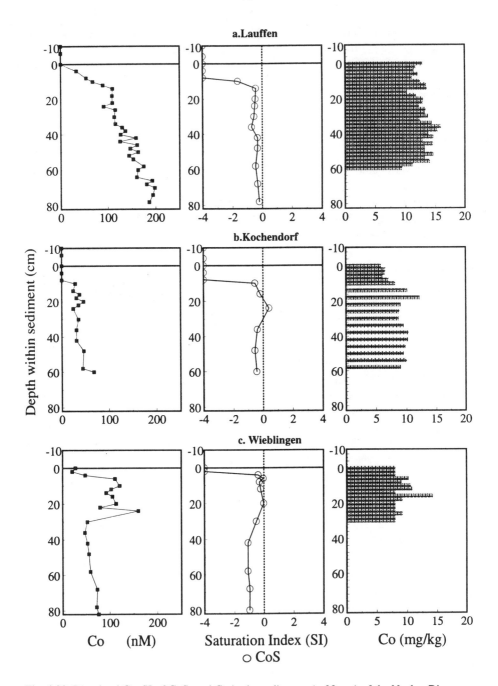

Fig. 4.32. Dissolved Co, SI of CoS, and Co in the sediments (< 20 μm) of the Neckar River

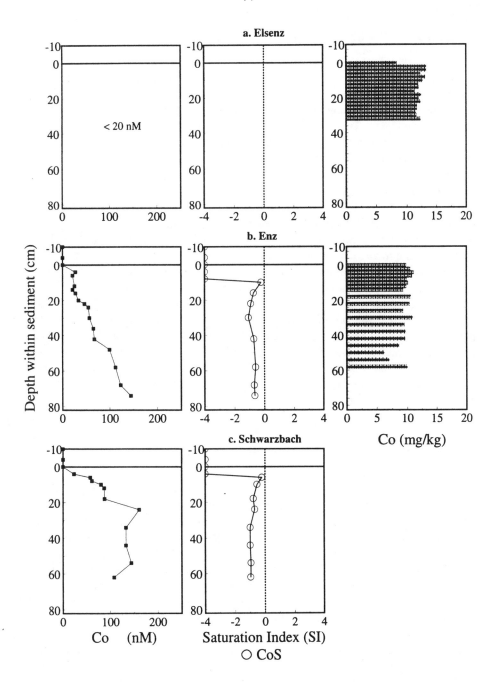

Fig. 4.33. Dissolved Co, SI of CoS, and Co in the sediments (< 20 μm) of the tributaries

4.3 Mobilization of heavy metals

Sediments are both carriers and potential sources of contaminants in aquatic systems. Natural and man-induced processes (e.g. flood, sediment erosion, dredging) may change physicochemical conditions in sediments, and consequently lead to the release of the contaminants. The large amounts of contaminated sediments in the Neckar River and its tributaries pose problems for this aquatic system. For instance, the resuspension and transport of sediments by flood occur often in the Neckar river. This may lead to a release of toxic substances from the sediments. Information about the mobilization of heavy metals in this area is necessary to evaluate whether heavy metals may be released into the overlying water and be bioavailable. This knowledge is essential for water management and sediment cleanup plans.

In surface waters, stable phases of heavy metals are adsorbed species. Fe and Mn oxides are the main carrier for Cd, Zn, and Ni, while the organic fraction is predominant for Cu and Pb (Tessier et al. 1985; Salomons et al. 1987). After deposition, the mineralization of organic matter may influence the main carriers. In this process, oxidation of organic matter, reduction of sulfate and Fe/Mn oxides are important. As reaction products, sulfide and organic colloids are released into the porewater, resulting in the formation of highly insoluble metal sulfide compounds and dissolved metal-organic complexes. There is a competition between metal-sulfide precipitation, metal-organic-complex dissolution, and Fe/Mn oxides adsorption. This competition can mainly control the mobility and bioavailability of heavy metals in sediments.

4.3.1 Metal-sulfide precipitation/dissolution and mobility of Cd, Zn, Pb, and Cu

On the basis of the solubility calculations for metal-sulfide compounds, heavy metals such as Cd, Zn, Pb, and Cu are supersaturated and remain fixed in the sediments as sulfide. Variations of redox potentials in sediments lead to changes in the binding forms of heavy metals on particles, which controls their chemical behavior in the sediments. If the sediments are removed from the reducing environment and have contact with O_2 in the overlying water, heavy metals bound to sulfides would be released by oxidation (Moore et al. 1988). In the suspension of anoxic sediments from the Hamburg harbor, pH values decreased from neutral (7.0) to strongly acidic (3.4) (Calmano et al. 1992). This may be due to the oxidation of sulfide:

$$FeS + 9/4\ O_2 + 5/2\ H_2O \rightarrow Fe(OH)_3 + SO_4^{2-} + 2\ H^+$$

Therefore, a low buffer capacity in the sediment samples cannot not prevent pH decrease. A significant release of heavy metals (Cd, Zn, Pb, and Cu) during the oxidation of anoxic sediments occurs only if the pH sinks below 4.5. Above this pH value, heavy metals dissolved due to the oxidation of anoxic sediments, may be adsorbed onto freshly formed Fe and Mn oxides surfaces. The pH is then a key factor for the mobility of heavy metals in sediments (Förstner et al. 1990; Calmano et al. 1992).

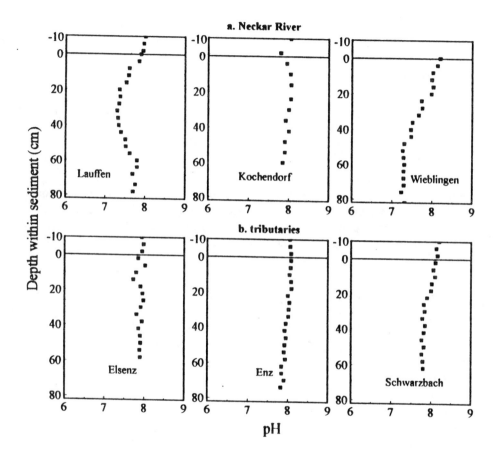

Fig. 4.34. The pH values in the porewater of the Neckar River and its tributaries

In the Neckar River and its tributaries, pH values varied from 7.3 to 8.2 (Fig. 4.34), they were always higher than 4.5. It is then necessary to assess the buffer capacity of the sediments.

For an estimation of sediments regarding their acid potential which can be produced by the oxidation of sulfide compounds, the Acid Producing Capacity (APC) and the Acid Neutralizing Capacity (ANC) are a useful tool. The important reactions affecting pH values in sediments are as follows (Calmano et al. 1992):

acid-producing reactions:

$$NH_4^+ + 2O_2 \rightarrow NO_3^- + H_2O + 2H^+$$
$$H_2S + 2O_2 \rightarrow SO_4^{2-} + 2H^+$$
$$R\text{-}SH + 2O_2 \rightarrow R\text{-}OH + SO_4^{2-} + 2H^+$$
$$S° + 3/2\ O_2 + H_2O \rightarrow SO_4^{2-} + 2H^+$$
$$FeS_2 + 15/4\ O_2 + 7/2\ H_2O \rightarrow Fe(OH)_3 + 2\ SO_4^{2-} + 4H^+$$
$$FeS + 9/4\ O_2 + 5/2\ H_2O \rightarrow Fe(OH)_3 + SO_4^{2-} + 2\ H^+$$
$$Fe^{2+} + 1/4\ O_2 + 5/2\ H_2O \rightarrow Fe(OH)_3 + 2H^+$$

acid neutralizing reactions (pH > 5):

$$HCO_3^- + H^+ \rightarrow CO_2 + H_2O$$
$$CaCO_3 + 2\ H^+ \rightarrow CO_2 + H_2O + Ca^{2+}$$

The ANC and the APC can then be calculated by

$$APC = APC_{aq} + APC_s$$
$$= \phi\ \Sigma\ f_i\ CP_{aq\ i} + (1\text{-}\phi)\ \rho_s\ \Sigma f_j\ CP_{s\ j}$$
$$ANC = ANC_{aq} + ANC_s$$
$$= \phi\ \Sigma f_i\ C^n_{aq\ i} + (1\text{-}\phi)\ \rho_s\ \Sigma\ f_j\ C^n_{s\ j}$$

where   f: stoichiometric acid producing (neutralizing) coefficient of the reaction,

        Caq: average dissolved concentrations in the uppermost 20 cm depth,

        Cs: average particulate concentrations in the uppermost 20 cm depth,

        ϕ : porosity of the sediments,

        ρ: density of the sediments.

Table 4.8. APC and ANC calculation of the sediments in the Neckar River at Lauffen

	C (mmol/1000cm^3)	f	APC mmol/kg	ANC mmol/kg
APCag-NH_4^+	1.2	2	1.8	
APCag-Fe^{2+}	0.10	2	0.1	
APCs-Sulfur	47	2	64	
ANCaq-HCO_3^-	11	2		16
ANCs-$CaCO_3$	1100	2		1500

Sulfur contents in the sediments were used to estimate the total concentrations of S°, FeS, FeS$_2$ and R-SH. Table 4.8 demonstrates that both APC and ANC are mainly controlled by solid phases. ANC in the sediments of the study area is much higher than APC, indicating a high buffer capacity of the sediments (Fig. 4.35). This is consistent with acid titration experiments of Neckar sediments by Kersten et al. (1985; Fig 4.36). The high buffer capacity of the sediments in the study area can be attributed to the high contents of carbonate (6.5-25%). Therefore, an oxidation of the anoxic sediments during resuspension must not lead to a decrease of pH values and a significant release of heavy metals into the overlying water.

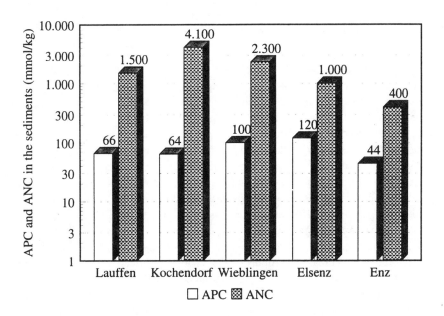

Fig. 4.35. APC and ANC calculated in the sediments of the Neckar River and its tributaries

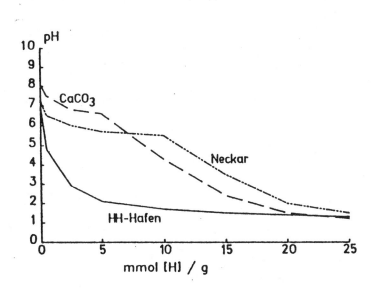

Fig.4.36. pH-titration curve for suspension of $CaCO_3$ and sediments of the Neckar River and the Elbe River (Hamburg harbor)

4.3.2 Influence of organic complexes and Fe/Mn redox processes

Natural and synthetic organic substances have been recognized as important factors in controlling transport and deposition of heavy metals in aquatic systems. The coexistence of organic ligands (e.g. humic substances) and metal ions in aquatic solutions often result in the formation of organic complexes. For example, humic acids can rapidly decompose sulfides such as sphalerite (ZnS), galena (PbS), or pyrite (Förstner and Wittmann 1979). Park and Huang (1989) reported that the dissolution of CdS is greatly enhanced in the presence of EDTA. As the mineralization of organic matter leads to the release of organic complexing materials into the porewater, dissolved organic carbon (DOC) increases with depth (Orem et al. 1986; Vuynovich 1989; Chin and Gschwend 1991). This increase may lead to the formation of metal-organic-complexes and the mobilization of heavy metals. Gerringa (1990) found that dissolved Cu concentrations are controlled by concentrations of organic ligands. The Cu concentrations decreased proportionally with the decrease of the ligand concentration. For Cd and Pb, no significant influence was found.

As organic complex formation enhances the mobility of heavy metals, it is important to know whether metals in the porewater are controlled by precipitation-dissolution reactions or by metal-organic-complex reactions. If complex formation is the dominating process, the increase of DOC should cause an increase in the concentrations of heavy metals. Indeed, dissolved Cd, Zn, Pb, and Cu did not increase with depth in the study area. Since the sediments are able to provide HS^- by the reduction of SO_4^{2-}, the precipitation of metal-sulfides, instead of the formation of organic omplexes, may control the concentrations of dissolved Cd, Zn, Pb, and Cu. Elderfield (1981) reported that 80 % of the dissolved Cu, 40 % of Fe were associated with organic material in the porewater from Narragansett Bay. The concentration of DOC increased with depth, but the concentrations of organic-bound metals decreased with depth. This can be interpreted that organic-bound metals are released from organic complexes and formed insoluble metal-sulfides. Similar results were reported by Douglas et al. (1986). It seems that the interaction of organics with metals is not sufficient to prevent the precipitation of metal-sulfides.

Another important factor influencing the mobility of heavy metals are Fe and Mn oxides. With the reduction of Fe and Mn oxides, heavy metals adsorbed on these oxides could be released into porewater. In a laboratory experiment, Francis and Dodeg (1990) found that there was a significant release of Cd, Pb, Zn, Ni, and Cr bound to Fe oxides during the anoxic degradation of organic matter.

It is then important to know whether dissolved metals are controlled by the adsorption/ desorption on Fe/Mn oxides, or by precipitation /dissolution of metal-sulfides in the sediments of the study area. If the letter is the case, the concentrations of heavy metals should be independent on the metal concentrations in the solid phase. However, an increase in the concentrations of dissolved Cd, Zn, Pb, and Cu with depth was not found in the study area. The probable explanation is that these metals released from the Fe and Mn oxides are rapidly precipitated as metal-sulfides. This is supported by the observation in the sediments of the Milltown Reservoir (Moore et al.

1988). They reported that Cu and Zn in the porewater are controlled by the solubility of Fe and Mn oxides in the oxic zone, and by metal-sulfides in the reduced zone.

Particulate Cr and Co are mainly bound to Fe/Mn oxides. With the reduction of these oxides, Cr and Co are simultaneously released into the porewater. Considering the concentration gradients, a diffusion of dissolved Fe, Mn, Cr, and Co towards the uppermost sediment layer is expected. In the surface layer, Fe^{2+} and Mn^{2+} were reoxidized as Fe/Mn oxides and immobilized. As a result, Cr and Co diffused from deeper sediment layer are scavenged by freshly formed Fe/Mn oxides. Therefore, Fe/Mn oxides in the surface layer act as a trap for Cr and Co in the sediments.

4.3.3 Degradation of organic matter and the mobility of heavy metals

The peaks of dissolved Cu and Zn at the sediment-water interface can be explained either as a result of leaching during the oxidation of metal-sulfides by oxygen in the overlying water, or as a result of the degradation of biomass which contains these metals. The seasonal variation of heavy metal profiles at the sediment-water interface at the site Lauffen is presented in Fig.4.37. Pronounced peaks of Cu and Zn are measured in the samples from June and October, a period of high sedimentation of biological materials. Large, rapidly sinking biogenic particles have been recognized as an important factor on the vertical transport of many elements in the water column. Evidence from sediment traps studies reflects that the plankton depositing to the sediments may comprise large particle fluxes of the elements, particularly following algal blooms (Hamilton-Tayler et al. 1984; Sigg et al. 1987; Morfett et al. 1988; Gerringa 1990; Balistrieri et al. 1992; Lee and Fisher 1992).

Sigg (1986) reported that mean elemental composition of the phytoplankton corresponds to the following stoichiometry:

$$(CH_2O)_{113}(NH_3)_{15}(H_3PO_4)Zn_{0.06}Cu_{0.008}Pb_{0.004}Cd_{0.00005}$$

This stoichiometry reflects the active biological uptake of Cu and Zn, essential elements for biota. Cd and Pb have no physiological functions, but an adsorption of Cd and Pb on biological surfaces is possible (Sigg et al. 1987). According to the relative affinity of trace metals to phytoplankton: $Zn \gg Cu > Pb \gg Cd$, the seasonal changes of Zn and Cu released by the decomposing phytoplankton should be measured at the surface sediments of the Neckar River. It is possible that small amounts of Pb and Cd released at the surface sediment layer may become adsorbed on other particles such as Fe/Mn oxides and clay minerals. Therefore, Pb and Cd in the porewater would be too low to be detected. Gendron et al. (1986) suggested that Cd in the Gulf of St. Lawrence is fixed by phytoplankton in the overlying water. The biological Cd is transported downward with the deposition of phytoplankton and subsequently released into the water during the aerobic degradation of organic matter. Similar results are reported by Morfett et al. (1988). In addition, the release of biogenic metals is confirmed by an incubation experiment. Gerringa (1990) found that dissolved Cu, Cd, Pb, and DOC increased rapidly after addition of easily degradable organic matter (shrimp). Table 4.9 compares the molar ratio of dissolved

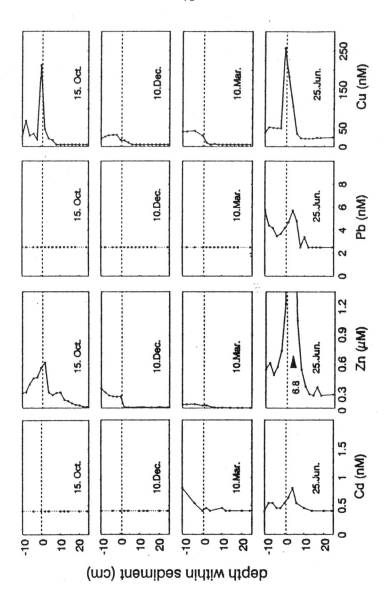

Fig. 4.37. Seasonal variation of heavy metals in the porewater at Lauffen

Zn and Cu at the sediment-water interface in the Neckar River. The good agreement with the stoichiometry given by Sigg (1986) suggests the biogenic source of dissolved Cu and Zn at the sediment-water interface.

Table 4.9. Dissolved Cu and Zn, and ratios of Cu and Zn at the sediment-water interface

	Zn (μM)	Cu (μM)	Zn:Cu
Lauffen	0.40	0.092	4.4
Kochendorf	0.27	0.051	5.3
Wieblingen	0.47	0.059	8.0
Biological stoichiometry			5-7.5

Sigg et al. (1987) found that the fluxes of Cu and Zn as well as those of Pb and Cd to the sediments of Lake Zurich are highest in summer (June-September) - the time of the highest sedimentation of biological material. Lee and Fisher (1992) reported that the release rate of Cd and Zn from the plankton cells is significantly higher at 18°C than that at 4°C, probably due to enhanced microbial activities. Likewise, the high concentrations of porewater Zn and Cu from June and October in the sediments at Lauffen again confirmed the importance of microbial activity on the release of heavy metals.

In summary, porewater profiles of Cd, Zn, Pb, and Cu reflect both the release of these metals during the decomposition of organic matter and their removal by precipitation or coprecipitation as metal-sulfides. The degradation of labile organic matter is most intensive at the sediment-water interface compared to the deeper sediments. Heavy metals released in the surface sediment layer diffuse both into the reducing zone where they are precipitated; and back into the overlying water, where they may be taken up by phytoplankton and recycled through biological processes.

4.4 Early diagenesis and heavy metal mobility in other freshwater Sediments: a comparison

4.4.1. Major rivers of Germany: Rhine, Main, Weser, and Elbe.

In the Rhine River sediments, Mn^{2+} varied between 0.55 - 4.7 µM in the bottom water, and increased rapidly below the sediment-water-interface. The maximum value (282 µM) was measured at 8 cm depth. This is 4 times higher compared that in the Neckar River Sediments, indicating more labile Mn oxide in the sediments. The decrease of Mn^{2+} may be explained as a result of precipitation of rhodochrosite. Low NO_3^- concentrations in the bottom water (0.44 mM) led to a slight NO_3^- reduction in the bottom water near the sediment-water-interface (Fig. 4.38). Fe^{2+} was released from the sediments into porewater below 0 cm depth.

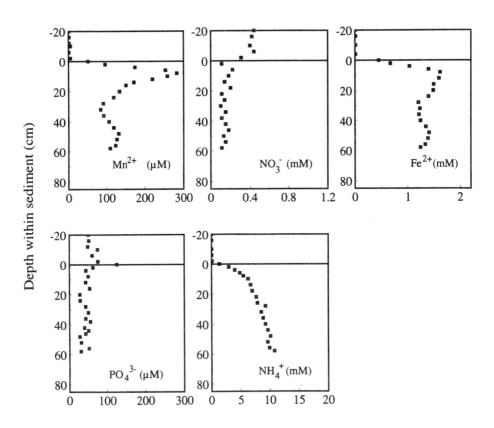

Fig. 4.38. Porewater profiles from the sediments of Rhine River (Wesel)

The SO_4^{2-} concentrations in the porewater were not measured. However, the steep concentration gradients of NH_4^+ between 0 and 10 cm depth indicates a strong SO_4^{2-} reduction process in this zone.

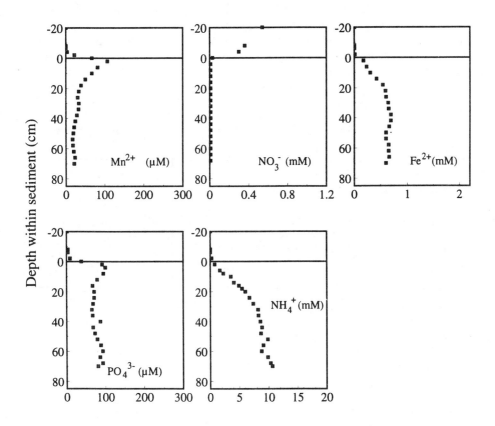

Fig. 4.39. Porewater profiles from the sediments of Main River (Erlabrunn)

Similar porewater profiles were found in the Main River sediments. Mn^{2+} was released into bottom water at 4 cm above the sediment-water-interface. The maximal value was only 107 µM at 2 cm depth. NO_3^- was completely consumed in the bottom water, so that only 0,01 mM NO_3^- was measured in the sediments. As Fe oxide was reduced with the oxidation of organic matter, Fe^{2+} increased with depth between 0 and 20 cm depth. PO_4^{3-} increased rapidly from 3.5 µM in the bottom water to 100 µM at 4 cm depth. As a product of the oxidation of organic matter, NH_4^+ increased continuously with depth (Fig. 4.39).

In Weser River sediment, Mn^{2+} and Fe^{2+} were released into porewater directly below the sediment-water-interface. The concentrations of Mn^{2+} and Fe^{2+} increased continuously with depth. The maximum values were 259 μM for Mn^{2+}, and 2,09 mM for Fe^{2+}, respectively (Fig. 4.40). High NO_3^- concentration (0.91 - 1.15 mM) in the bottom water resulted in a rapidly NO_3^- reduction process. Similar to NH_4^+ profiles in the Neckar River sediments, steep NH_4^+ concentration gradients between 0 and 30 cm depth reflects a strong SO_4^{2-} reduction process in this zone. Below 30 cm depth, NH_4^+ concentration varied between 11.5 and 12.7 mM. Similar to NH_4^+ profiles, PO_4^{2-} concentrations increased continuously between 0 and 30 cm depth, and were constant (about 60 μM) below 30 cm depth.

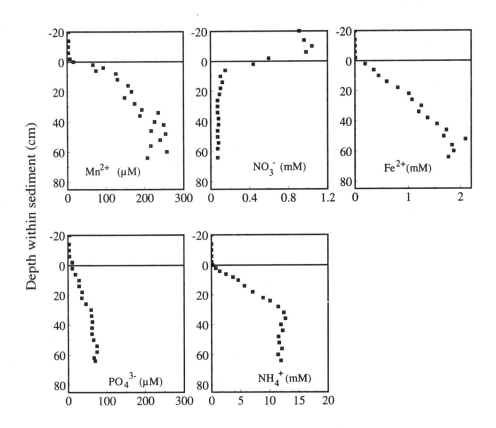

Fig. 4.40. Porewater profiles in the sediments of the Weser River (Bremen)

In the Elbe River sediments, a significant release of Mn^{2+} into the porewater occurred below 16 cm depth. NO_3^- concentrations varied between 0.54 and 0.91 mM above 52 cm depth. The reason of the high NO_3^- concentrations (2.2-3.5 mM) between

54 and 62 cm depth is not clear (Fig. 4.41). Fe oxides was reduced below 20 cm depth. NH_4^+ concentrations increased rapidly from 0.50 mM at 16 cm depth to 17 mM at 32 cm depth, indicating a strong SO_4^{2-} reduction in this zone. Low concentrations of Mn^{2+}, Fe^{2+}, and NH_4^+ between 0 and 16 cm depth reflect the layer of O_2 penetration.

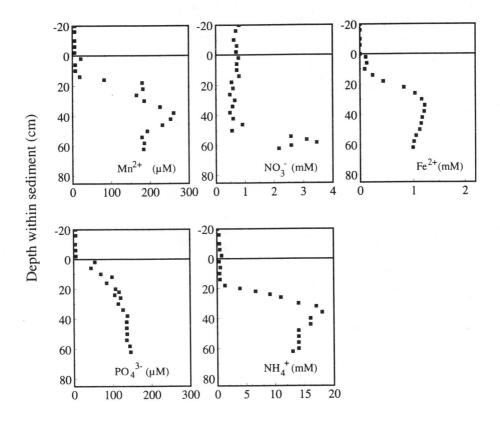

4.41. Porewater profiles in the sediments of the Elbe River (Hamburg)

In sediments of the major rivers, porewater profiles of NO_3^-, Mn^{2+}, Fe^{2+}, PO_4^{3-}, and NH_4^+ are typical of anoxic organic-rich sediments. High input of organic matter in the sediments of the major rivers are confirmed by the contents of organic carbon of the sediments. Fig. 4.42).

We did not measure O_2 concentrations in the porewater. However, Mn^{2+} and NO_3^- profiles reflect a complete O_2-consumption in the bottom water. Below the sediment-water-interface, SO_4^{2-} reduction and methane fermentation are the dominating redox reactions, indicating a strong anoxic condition in the sediments. This play an important role in the mobility of nutrients and heavy metals. For example, the concentration of PO_4^{3-} in the sediments was probably controlled by the formation of vivianite. PO_4^{3-} diffused into the oxic bottom water would be adsorbed by fresh formed Fe oxide.

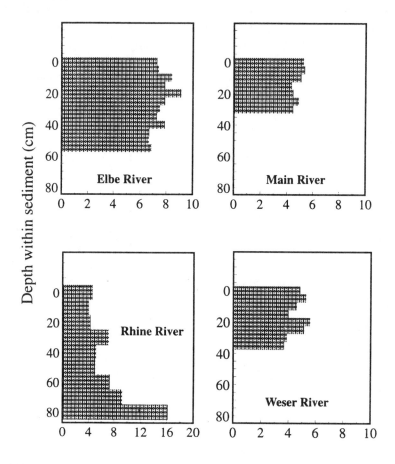

Fig. 4.42. The contents of organic carbon (Corg, %) in the sediments of major rivers.

Fig. 4.43 shows distribution of Cu between sediments and porewater in the sediments of major rivers. In the Rhine River sediments (Wesel), particulate Cu decreased from 430 mg/kg at 80 cm depth to 120 mg/kg at 50 cm depth. As reported by Förstner and Müller (1974), high concentrations of Cu in the deeper sediments (below 50 cm depth) can be attributed to the industrial activities before 1970. In the Elbe River sediments (Hamburg), a decrease of Cu concentrations occurred at 15 cm depth. Lowest concentrations of particulate Cu (69-88 mg/kg) were found in the Weser River sediments (Bremen).

Porewater Cu showed a different tendency. In the Weser River, dissolved Cu decreased from 75 nM in the bottom water to 5.4 nM at 64 cm depth. Similar profiles are also found in the Elbe River and the Main River. In Rhine River, where the concentrations of particulate Cu were the highest, dissolved Cu in the most of porewater samples were lower than 10 nM..

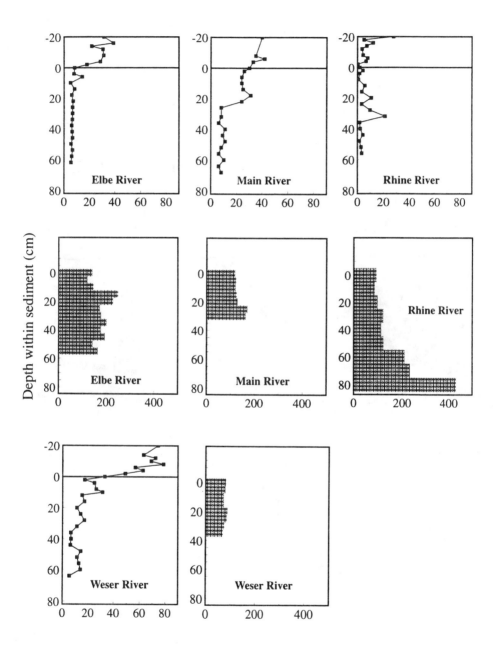

Fig. 4.43. Dissolved Cu (▬■▬ , nM) and particulate Cu (▦ , mg/kg) in the sediments

Highest concentrations of particulate Zn (3200 mg/kg) was found in the sediments of Rhine River at 80 cm depth. The concentrations of particulate Zn decreased to 840 mg/kg at 50 cm depth (Fig. 4.44).

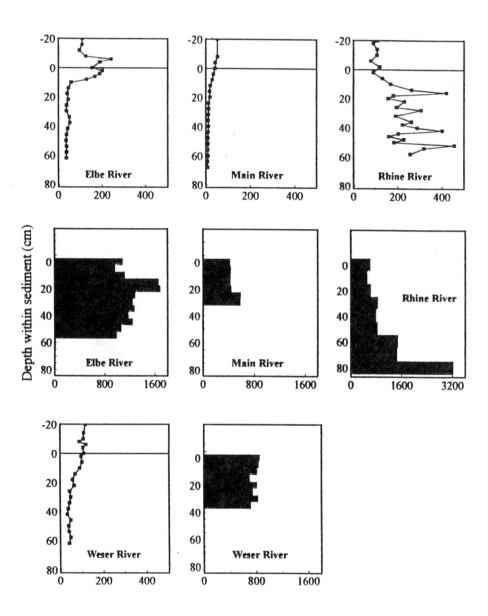

Fig. 4.44. Dissolved Zn (─■─ , nM) and particulate Zn (▬▬ , mg/kg) in the sediments.

In the Elbe, Main, and Weser River, the concentrations of dissolved Zn in the porewater was lower compared to the overlying water. In the Rhine River sediments, porewater Zn increased with depth.

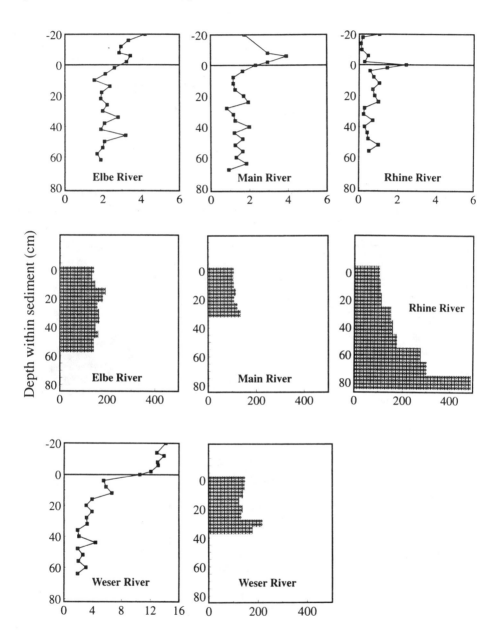

Fig. 4.45. Dissolved Pb (—■— , nM) and particulate Pb (▦ , mg/kg) in the sediments

In the Rhine River sediments, the concentrations of particulate Pb increased from 110 mg/kg at 6 cm depth to 490 mg/kg at 80 cm depth. In other river sediments, the concentrations of particulate Pb varied between 100 and 220 mg/kg (Fig. 4.45).

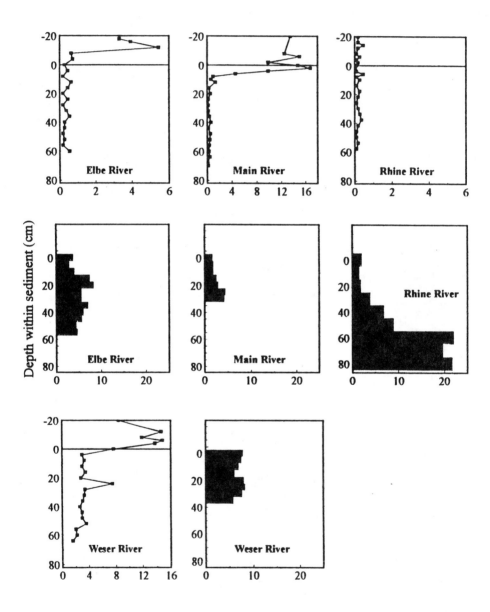

Fig. 4.46. Dissolved Cd (▬■▬ , nM) and particulate Cd (▬▬ , mg/kg) in the sediments.

In the Rhine River and Main River, Notable peaks of dissolved Pb were measured at the sediment-water interface. In the Elbe River and Weser River, the concentrations of porewater Zn were lower than compared to the overlying water.

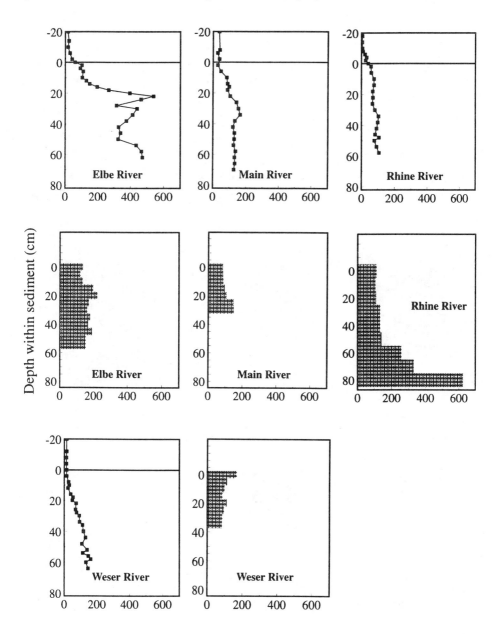

Fig. 4.47. Dissolved Cr (▬■▬ , nM) and particulate Cr (▦ , mg/kg) in the sediments

In the Rhine River sediments, the concentrations of particulate Cd were 21 mg/kg between 60 and 80 cm depth, 9 times higher compared to the surface layer. In other river sediments, Cd concentrations were lower than 10 mg/kg (Fig. 4.46).

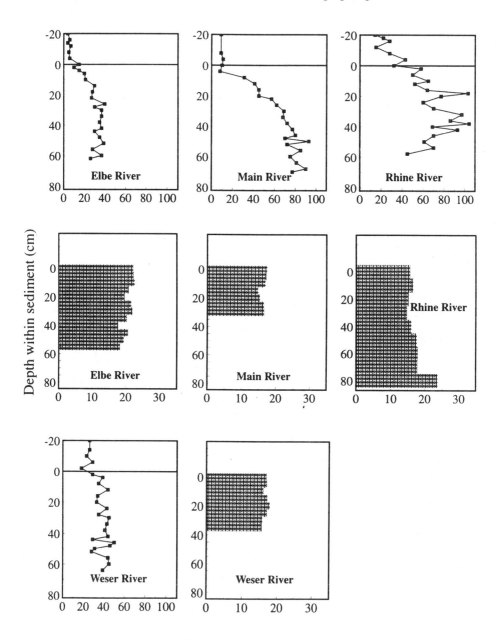

Fig. 4.48. Dissolved Co (▬■▬ , nM) and particulate Co (▦ , mg/kg) in the sediments

In the Elbe River, Dissolved Cd decreased from 5.4 nM in the overlying water to 0.4 nM near the sediment-water interface. In the Main River, Such a decrease occurred between 0 and 8 cm depth. The concentrations of dissolved Cd were lower than 0.4 nM in the Rhine River sediments, even in the presence of elevated paticulate Cd concentrations.

In contrast to the porewater profiles of Cu, Zn, Pb, and Cd, dissolved Cr and Co increased with depth (Fig. 4.47, 4.48).

In summary, the concentrations of heavy metals in the porewater were independent of those in the sediments. A general downward increase of the concentrations of particulate Cd, Zn, Pb, Cu can be observed. In contrast, a downward decrease of the concentrations of dissolved Cd, Zn, Pb, and Cu was found. As reported for the Neckar River sediments, the decrease of porewater Cu, Zn, Pb, and Cd with depth can be attributed to the formation of metal-sulfides. High contents of carbonate of the sediments may prevent an decrease of pH value in the porewater and the overlying water (Fig. 4.49). A significant release of heavy metals during resuspension of the sediments cannot be expected.

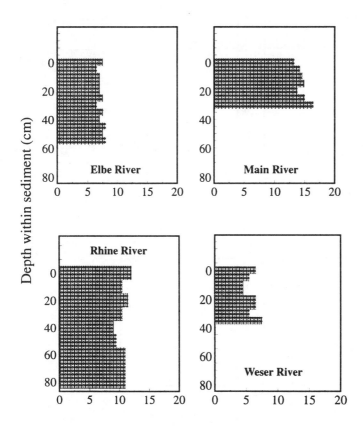

Fig. 4.49. Carbonate contents (%) in the sediments of the major rivers.

4.4.2 Lake Constance (Uttwil)

Organic carbon in the sediments of the Lake Constance was only 2.5% compared to

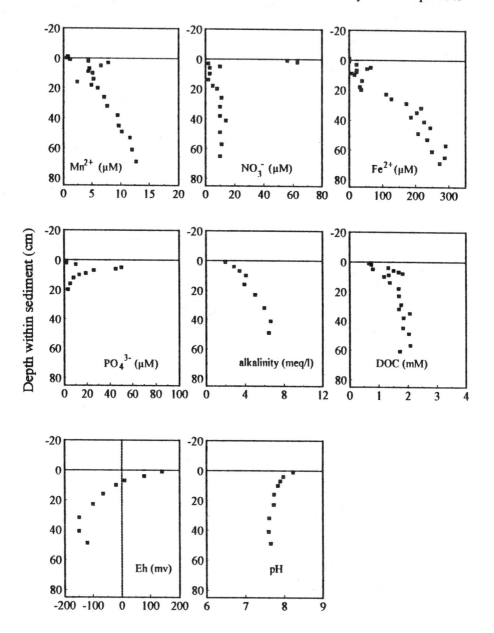

Fig. 4.50. Porewater profiles in the sediments of Lake Constance (Uttwil)

4 % in the Neckar River sediments, indicating low input of organic matter to the sediment as a result, little concentration gradients were found in the sediments (Fig. 4.50). Mn^{2+} was released below the sediment-water-interface. NO_3^- in the overlying water was only 0.056 mM compared to 0.40 mM in the Neckar River water. Fe^{2+} increased from 36 μM at 20 cm depth to 290 μM at 65 cm depth. The maximum value of PO_4^{3-} (50 μM) was measured at 5 cm depth. As an indicator of the mineralization of organic matter, alkalinity increased with depth. The maximum value was only 6.6 meq/l compared to 43 meq/l in the Neckar River sediments. Dissolved organic carbon (DOC) increased from 0.67 mM in the supernating water to 2.1 mM at 57 cm depth. This can be attributed to the decomposition of organic matter. In addition, Eh values were negative below 10 cm depth, reflecting O_2-consumption by the mineralization of organic matter.

4.4.3. Mobility of heavy metals in Lean River, P. R. China

In the Lean River sediments (Caijiawan), NO_3^- concentrations decreased from 0.045 mM in the overlying water to 0.009 mM at 34 cm depth (Fig. 4.51). This profile was similar to that in Lake Constance. However, the concentrations of NO_3^- in the bottom water were much lower than those in the Neckar River (0.42 mM), indicating a weak NO_3^- reduction process. The concentrations of Mn^{2+} increased with depth between 0 and 26 cm depth. Fe^{2+} was released into the porewater below 6 cm depth. As a product of the mineralization of organic matter, NH_4^+ increased with depth. It is notable that the maximum value of NH_4^+ was only 0.47 mM compared to 3 - 15 mM in the Neckar River sediments.

Organic carbon content in the sediments of Lean River at Caijiawan was 0.87 %. Porewater profiles in this river are typical of organic-poor sediments. The profiles of Mn^{2+}, Fe^{2+}, NH_4^+, and SO_4^{2-} in the sediments reflect anoxic conditions in the sediments. This may cause the formation of metal-sufides in the sediments, and consequently immobility of heavy metals in this aquatic systems.

Fig. 4.52 shows distribution of heavy metals between sediments and porewater. No significant relationship was found in the concentrations between porewater and sediments. The concentrations of Cu in the sediments were very high (about 1000 mg/kg), the concentrations of dissolved Cu were lower than 200 nM. Dissolved Pb was below the detection limit (2.5 nM).

As in other anoxic freshwater sediments, transport of heavy metals from the dissolved to the particulate phase can be explained by the precipitation of metal-sulfides. As the carbonate concentrations of the sediments were very low (< 0.2%), the acid neutralizing capacity of the sediments is low. Therefore, whether the heavy metals might be released into the overlying water by resuspension of the sediments, remains an open question.

94

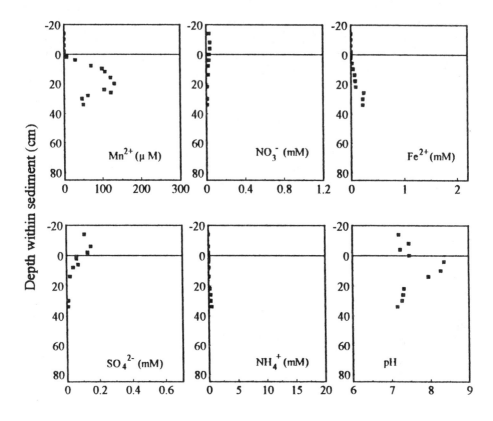

Fig. 4.51. Porewater profiles in the sediments of the Lean River (Caijiawan, China)

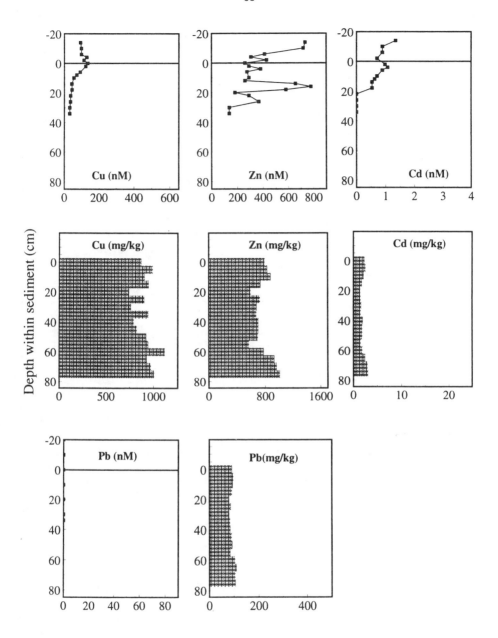

Fig. 4.52. Distributions of heavy metals in the sediments and porewater in the Lean River (Caijiawan, China)

4.4.4 Early diagenesis in the Moscow and Oka River (Russia)

Fig. 4.53, 4.54 show the porewater profiles from the Moscow River and the Oka River.

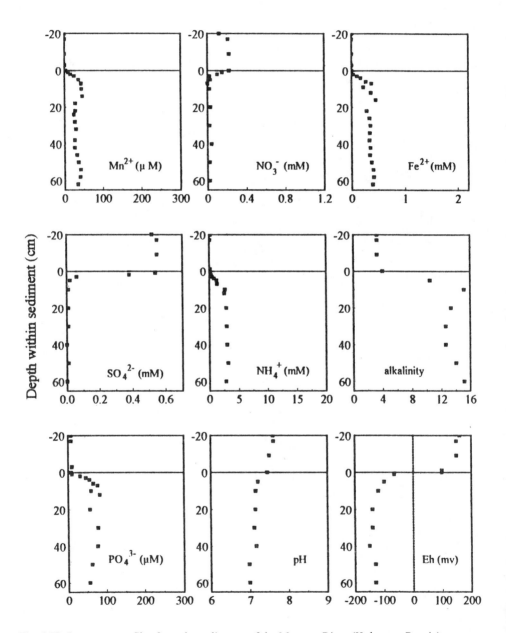

Fig. 4.53. Porewater profiles from the sediments of the Moscow River (Kolomna, Russia)

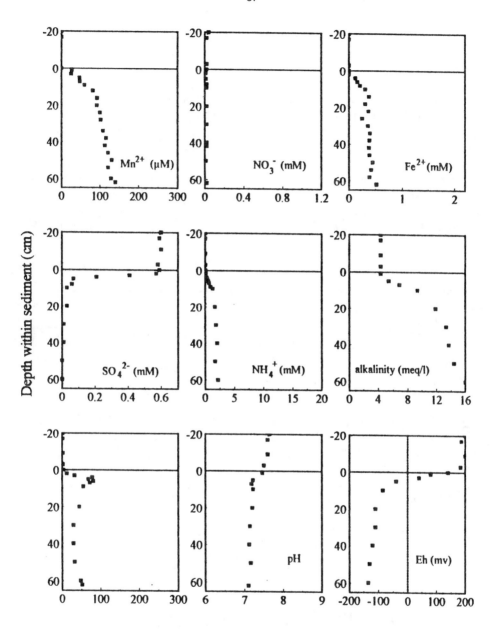

Fig. 4.54. Porewater profiles from the sediments of the Oka River (Kolomna, Russia)

The Oka River mainly drains area agricultural area. However, the Moscow River, one of the tributaries of the Oka River, flows through the most important industrial region in Russia. As a result, the Moscow River is moderately to strongly polluted with heavy metals (Cu, Zn, Pb, and Cr). After joining the Oka River at Kolomna, the Oka sediments clearly showthe negative influence of the Moscow River.

The porewater profiles in the sediments of both rivers show an anoxic conditions in the sediments (Fig. 4.53, 4.54). In the Moscow River, Eh values decreased from 160 mV in the overlying water to -65 mV at 1 cm depth. In the Oka River, Eh values decreased from 200 mV in the overlying water to -40 mV at 5 cm depth. This can be attributed to the consumption of O_2 by the mineralization of organic matter. In this process, Mn oxide and Fe oxide were reduced, and Mn^{2+} and Fe^{2+} were released into the porewater. The maximum value of Mn^{2+} in the Oka River sediments was 140 µM compared to 48 µM in the Moscow River sediments, reflecting more labile Mn-

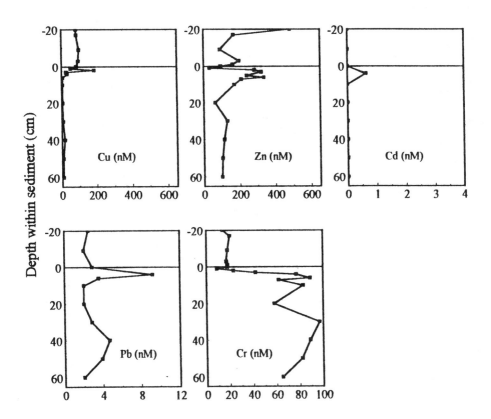

Fig. 4.55. Distribution of dissolved heavy metals in the Moscow River sediments(Kolomna)

oxides in the Oka River sediments. As products of the decomposition of organic matter, PO_4^{3-} and NH_4^+ were released into the porewater. The maximum values of NH_4^+ were 3.0 mM and 2.2 mM in the sediments of Moscow River and Oka River, respectively. These values are lower compared to the Neckar River sediments (4 - 15 mM). This is in good agreement with the low organic carbon content (2.5%) in the sediments of the Moscow River and Oka River.

In the sediments of the both river, notable peaks of the dissolved Cu, Zn, Pb, and Cd were measured at the sediment-water-interface (Fig. 4.55, 4.56). Cu and Zn have a physiological function in biomas. In addition, Pb and Cd can be adsorbed by biomasss. Therefore, the peaks of dissolved heavy metals can be interpreted by decomposition of the organic materials at the surface sediment layer, which contained these heavy metals. With the reduction of Fe oxide and Mn oxide, dissolved Cr was released into the porewater.

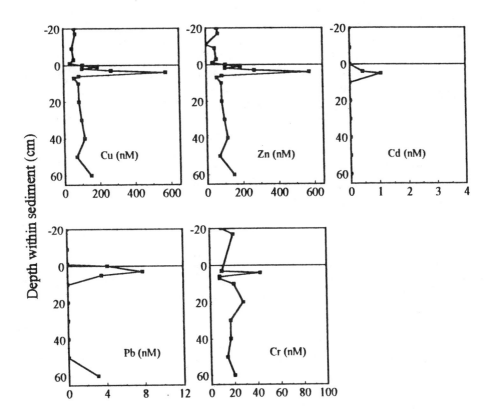

Fig. 4.56. Distribution of dissolved heavy metals in the Oka River sediments (Kolomna)

4.5 Comparison of the early diagenesis in freshwater and marine sediments

Organic matter deposited in freshwater and marine sediments is usually reduced by O_2, Mn oxide, NO_3^-, Fe oxide and SO_4^{2-}. After consumption of these oxidants, methane fermentation is a dominating process. However, different processes dominate mineralization of organic matter in each system. This may be attributed to differences in ionic composition and input of different organic matter in each system.

Quality and quantity of organic matter play an important role in the mineralization of organic substances. The more labile components reach the sediment, the faster mineralization processes take place. The flux of organic matter depends on the primary productivity in the supernating water. As much more nutrients are discharged into freshwater compared to marine systems, most freshwater sediments have high organic carbon contents. As a result, O_2 is rapidly consumed in the surface layer in freshwater sediments, anoxic conditions dominate these sediments (Jørgensen 1983; Carignan and Nriagu 1985; Schwedhelm et al. 1988; Vuynovich 1989; Williams 1992). In contrast, marine sediments have normally low organic carbon contents. Furthermore, labile organic particles in the deep sea are usually oxidized by O_2 in seawater before they reach the sediment surface. The depth of O_2 penetration is several centimeters compared to a few millimeter in freshwater sediments.

SO_4^{2-} reduction is a dominant anoxic reduction process in marine sediments, whereas low sulfate concentrations generally limit the rate of sulfate reduction in freshwater sediments. Because of the rapid consumption of SO_4^{2-} in freshwater sediments (between 0 and 20 cm depth, major rivers of Germany), methane fermentation plays an important role in the mineralization processes.

Because of low concentrations of NO_3^- in marine sediments, NO_3^- reduction does not play a significant role in the mineralization of organic matter (Balzer 1989). However, the research both in marine (Sagemann et al 1994) and in freshwater sediments (Neckar River) show the relationship between denitrification rate and temperature, resulting from bacterial activity.

In marine sediments, the zones of aerobic oxidation, NO_3^- reduction, Mn oxide reduction, Fe oxide reduction, and SO_4^{2-} reduction can be spatially separated (Froelich et al. 1979). This is not true in freshwater sediments because of the rapid decomposition of organic matter.

Marine sediments are a site of decomposition and nutrient regeneration from organic matter. They provide essential nutrients for biomass formation and play an important role in cycling of C, N, and P in the ocean. In freshwater environments, the rapid decomposition of organic matter in anoxic sediments leads to a strong release of NH_4^+ and PO_4^{3-} into the porewater. High concentrations of NH_4^+ (2 - 16 mM) were measured in freshwater sediments compared to 0.03 mM in marine sediments (Froelich et al. 1979). A large scale release of NH_4^+ and Fe^{2+} into the supernating water could cause an O_2-depletion, which in extreme cases could result in an ecological catastrophe.

5 Conclusions

A series of geochemical profiles in both porewater and solid phases in the freshwater sediments of major rivers in Germany (Neckar, Rhine, Elbe, Main, and Weser), Lake Constance, Lean River (China), Moscow River and Oka River (Russia) lead to the conclusion, that the mineralization of organic matter plays an important role in the cycling of nutrients and heavy metals. First, heavy metals and nutrients bound to organic particles (organic C, N, P) are released in inorganic forms (CO_2, PO_4^{3-}, NH_4^+, and heavy metal ions) into the porewater. They either diffuse into overlying water or are precipitated with other ions, depending on the physicochemical conditions. Second, the mineralization processes cause a change of redox potential, adsorption capacity of solid phases (Fe oxide and Mn oxide), and concentration of organic complexes. This can change binding forms of the nutrients and heavy metals on particles, and indirectly influence their mobility.

Because of high input of organic material, anoxic conditions are dominating in freshwater sediments. Low concentrations of dissolved Cu, Zn, Pb, and Cd in the sediments, even in heavily polluted sediments (Neckar, Rhine, and Lean), are associated with the precipitation of metal-sulfides. In the sediments of Neckar, Rhine, Main, Weser, and Elbe, an acidification of the sediments during resuspension of the sediments cannot be expected due to high carbonate contents (8-15 %) of the sediments. In low acid neutralizing sediments such as from the Lean River (carbonate contents < 2%), it is an open question whether heavy metals are released into the supernating water. Furthermore, in sediments having low acid neutralizing capacity, dredging and on-site or off-site disposal, would minimize the controls imposed by the reducing environment and mobilize metals into surface and groundwater systems. The best management for such systems is the stabilization of the structure retaining the sediments in their present condition and to minimize the variations in oxic-anoxic conditions in the sediments resulting from changes in water discharge.

6 References

Adams, W.J., Kimerle, R.A:, and Barnett, J.W.Jr. 1992. Sediment quality and aquatic life assessment. Environ. Sci. Technol. 26(10): 1865-1875.

Aller, R.C. 1980a. Diagenetic processes near the sediment-water interface of Long Island Sound. I. Decomposition and nutrient element geochemistry (S, N, P). Advances in Geophysics. 22:237-350.

Aller, R.C. 1980b. Diagenetic processes near the sediment-water interface of Long Island Sound. II. Fe and Mn. Advances in Geophysics. 22:350-415.

Balistrieri, L.S., Murray, J.W., and Paul, B. 1992. The biogeochemical cycling of trace metals in the water column of Lake Sammamish, Washington: Response to seasonally anoxic conditions. Limnol. Oceangr. 37(3): 529-548.

Balzer,W.1984. Organic matter degradation and biogenic element cycling in a nearshore sediment (Kiel Bight). Limnol. Oceanogr. 29(6): 1231-1246.

Balzer, W. 1989. Chemische Reaktionen und Transportprozesse in oberflächennahen Sedimenten borealer und polarer Meeresgebiete. Habilitationsschrift, Universität Kiel. 312 S.

Barbanti, A., Ceccherelli, V.C., Frascari, F., Reggiani, G., and Rosso, G. 1992a. Nutrient regeneration process in bottom sediments in a Po delta lagoon (Italy) and the role of bioturbation in determining the fluxes at the sediment-water interface. Hydrobiologia. 228: 1-21.

Barbanti, A., Ceccherelli, V.C., Frascari, F., Rosso, G., and Reggiani, G. 1992b. Nutrient release from sediments and the role of bioturbation in the Goro Lagoon (Italy). Science of the Total Environment. Supplement 1992: 475-487.

Baur, W. 1980. Gewässergüte. Verlag Paul Parey. Hamburg und Berlin. 144S.

Bender, M.L., Fanning, K.A., Froelich, P.N., and Maynard, V. 1977. Interstitial nitrate profiles and oxidation of sedimentary organic matter in the Eastern Equatorial Atlantic. Science. 198:605-609.

Bender, M.L. and Heggie, D.T. 1984. Fate of organic carbon reaching the deep sea floor: a status report. Geochim. Cosmochim. Acta. 48: 977-986.

Berner, R.A. 1980. Early diagenesis. Princeton University Press, Princeton, N.J. 241 S.

Billen, G. 1982. Modelling the processes of organic matter degradation and nutrients recycling in sedimentary systems. in Sediment Microbiology. edited by Nedwell, D.B. and Brown, C.M. Academia Press, London.

Boers, P. and Bles, F. 1991. Ion concentrations in interstitial water as indicators for phosphorus release processes and reactions. Wat. Res. 25(5): 591-598.

Boudreau, B.P., Canfield, D.E., and Mucci, A. 1992. Early diagenesis in a marine sapropel, Mangrove Lake, Bermuda. Limnol. Oceanogr. 37(8): 1738-1753.

Boulegue, J., Lord III. C.J., and Church, T.M. 1982. Sulfur speciation and associated trace metals (Fe, Cu) in the pore waters of Great Marsh, Delaware. Geochim. Cosmochim. Acta. 46: 453-464.

Bruland, K.W., Donat, J.R., and Hutchins, D.A. 1991. Interactive influences of bioactive trace metals on biological production in oceanic waters. Limnol. Oceangr. 36(8): 1555-1577.

Calmano, W., J. Hong, and U. Förstner. 1992. Influence of pH value and redox potential on binding and mobilization of heavy metals in contaminated sediments. Vom Wasser. 78: 245-257.

Calvert, S.E. and Karlin, R.E. 1991. Relationships between sulphur, organic carbon, and iron in the modern sediments of the Black Sea. Geochim. Cosmochim. Acta. 55: 2483-2490.

Capone, D.G., Reese, D.D., and Kiene, R.P. 1983. Effects of metals on methanogenesis, sulfate reduction, carbon dioxide evolution, and microbial biomass in anoxic salt marsh sediments. Appli. Environ. Microbiol. 45:1586-1591.

Capone, D.G. and Kiene, R.P. 1988. Comparison of microbial dynamics in marine and freshwater sediments: Contrasts in anaerobic carbon catabolism. Limnol. Oceanogr. 33(4): 725-749.

Carignan, R. 1984. Interstitial water sampling by diagenesis: Methodological notes. Limnol. Oceangr. 29(3): 667-670.

Carignan, R. 1985. Quantitative importance of alkalinity flux from the sediments of acid lakes. Nature. 317: 158-160.

Carignan, R and Nriagu, J.O. 1985. Trace metal deposition and mobility in the sediments of two lakes near Sudbury, Ontario. Geochim. Cosmochim. Acta. 49: 1753-1764.

Carignan, R. and Tessier, A. 1985. Zinc deposition in acid lakes: the role of diffusion. Science. 228: 1524-1526.

Carignan, R. and Tessier, A. 1988. The co-diagenesis of sulfur and iron in acid lake sediments of southwestern Quebec. Geochim. Cosmochim. Acta. 52: 1179-1188.

Carignan, R. and Lean, D.R.S. 1991. Regeneration of dissolved substances in a seasonally anoxic lake: The relative importance of processes occurring in the water column and in the sediments. Limnol. Oceanogr. 36(43): 683-707.

Chanton, J.P. and Martens, C.S. 1987. Biogeochemical cycling in organo-rich coastal marine basin. 7. Sulfur mass balance, oxygen uptake and sulfide retention. Geochim. Cosmochim. Acta. 51: 1187-1199.

Chin, Y. and Gschwend, P.M. 1991. The abundance, distribution, and configuration of porewater organic colloids in recent sediments. Geochim. Cosmochim. Acta. 55: 1309-1317.

Cooke, J.M. and White, R.E. 1988. Nitrate enhancement of nitrification depth in sediment/water microcosms. Environ. Geol. Water Sci. 11(1): 85-94.

Cranston , R.E. and Murray, J.W. 1980. Chromium species in the North Pacific. Earth Planet.Sci. Letter. 47:176-198.

Dahmke, A., H.D. Schulz, and W. Weber. 1986. Mineralstabilitäten und Frühdiagenese in Ostseesedimenten. Meyniana, 38:109-124.

Dahmke, A, H.D. Schulz, A. Kölling, F. Kracht, and A. Lücke. 1991. Schwermetallspuren und geochemische Gleichgewichte zwischen Porenlösung und Sediment im Wesermündungsgebiet. Berichte, Fachbereich Geowissenschaften, Universität Bremen, Nr. 12.

Davison, W., Woof, C., and Rigg, E. 1982. The dynamics of iron and manganese in a seasonally anoxic lake; direct measurement of fluxes using sediment traps. Limnol. Oceanogr. 27(6): 987-1003.

Davison, W. 1986. Conceptual models for transport at a redox boundary. in Chemical Processes in Lakes. Edited by Stumm, W. John Wiley & Sons.

Deutsche Kommission zur Reinhaltung des Rheins. 1991. Zahlentafeln der physikalisch-chemischen Untersuchungen 1990.

Douglas, G.S., Mills, G.L., and Quinn, J.G. 1986. Organic copper and chromium complexes in the interstitial waters of Narragansett Bay sediments. Marine Chemistry. 19: 161-174.

Eck, G.T.M. van and Smits, J.G.C. 1986. Calculation of nutrient fluxes across the sediment-water interface in shallow lakes. in Sediments and Water Interactions. edited by Sly, P.G. Springer-Verlag.

Elderfield, H. 1981. Metal-organic associations in interstitial waters of Narragansett Bay sediments. American Journal of Science. 281: 1184-1196.

Elderfield, H., Mccaffrey, R.J., Luedtke, N., Bender, M., and Truesdale, V.W. 1981. Chemical Diagenesis in Narragansett Bay sediments. American Journal of Science. 281:1021-1055.

Elsgaard, L. and Jørgensen, B.B. 1992. Anoxic transformations of radiolabeled hydrogen sulfide in marine and freshwater sediments. Geochim. Cosmochim. Acta. 56: 2425-2435.

Emerson, S. 1976. Early diagenesis in anaerobic lake sediments: chemical equilibria in interstitial waters. Geochimica et Coschimica Acta. 40: 925-934.

Fernex, F.E., D. Span, G.N. Flatau, and D. Renard. 1986. Behavior of some metals in surficial sediments of the Northwest Mediteranean Continental Shelf. in Sediment and Water Interactions, edited by Sly, P.G. Springer-Verlag.

Förstner, U. and Müller, G. 1973. Anorganische Schadstoffe im Neckar. Ruperto Carola. 51: 67-71.

Förstner, U. and Müller, G. 1974. Schwermetalle in Flüssen und Seen. Springer-Verlag.

Förstner, U. and Wittmann, F. 1979. Metal pollution in the aquatic environment. Springer-Verlag. 486S.

Förstner, U., Ahlf, W., Calmano, W., Kersten, M. 1990. Sediment criteria development. In Sediments and environmental geochemistry. Edited by Heling, D., Rother, P., Förstner, U., and Stoffers, P. Springer Verlag.

Fossing, H. and Jørgensen, B.B. 1990. Oxidation and reduction of radiolabeled inorganic sulfur compounds in an estuarine sediment, Kysing Fjord, Denmark. Geochim. Cosmochim. Acta. 54: 2731-2742.

Francis, A.J. and Dodge, C.J. 1990. Anaerobic microbial remobilization of toxic metals coprecipitated with iron oxide. Environ. Sci. Technol. 24(3): 373-378.

Froelich, F.N., G.P. Klinkhammer, M.L. Bender, N.A. Luedtke, G.R. Heath, D. Cullen, P. Dauphin, D. Hammond, B. Hartman, and V. Maynard. 1979. Early oxidation of organic matter in pelagic sediments of the eastern equatorial Atlantic: suboxic diagenesis. Geochim. Cosmochim. Acta. 43: 1075-1090.

Gambrell, R.P., Wiesepape, J.B., Patrick, W.H., and Duff, M.C. 1991. The effects of pH, and salinity on metal release from a contaminated sediment. Water, Air, and Soil Pollution. 57-58: 359-367.

Gendron, A., Silverberg, N., Sundby, B., and Lebel, J. 1986. Early diagenesis of cadmium and cobalt in sediments of the Laurentian Trough. Geochim. Cosmochim. Acta. 50: 741-747.

Gerringa, L.J.A. 1990. Aerobic degradation of organic matter and the mobility of Cu, Cd, Ni, Pb, Zn, Fe, and Mn in marine sediment slurries. Marine Chemistry. 29: 355-374

Gobeil, C., Silverberg, N., Sundby, B., and Cossa, D. 1987. Cadmium diagenesis in Laurentian Trough sediments. Geochim. Cosmochim. Acta. 51: 589-596.

Goloway, F. and Bender, M. 1982. Diagenetic models of interstitial nitrate profiles in deep sea suboxic sediments. Limnol. Oceanogr. 27(4): 624-638.

Gschwend, P.M., MacFarlane, J.K., Newman, K.A. 1985. Volatile halogenated organic compounds released to seawater from temperate marine macroalage. Science. 227:1033-1035.

Gunnarsson, L.A.H. and Rönnow, P.H. 1982. Interrelationships between sulfate reducing and methane producing bacteria in coastal sediments with intense sulfide production. 1982. Marine Biology. 69:121-128.

Hamilton-Taylor, J. Millis, M., and Reynolds, C.S. 1984. Depositional fluxes of metals and phytoplankton in Windermere as measured by sediment traps. Limnol. Oceangr. 29(4): 695-710.

Hesslein, R.H. 1976. An in situ sampler for close interval pore water studies. Limnol. Oceangr. 21: 912-915.

Holdren, G.C. and Armstrong, D.E. 1986. Interstitial ion concentrations as an indicator of phosphorus release and mineral formation in lake sediments. In: Sediments and Water Interactions, edited by Sly, P.G. Springer-Verlag.

Huynh-Ngoc, L., Whitehead, N.E., Boussemart, M., and Calmet, D. 1989. Dissolved nickel and cobalt in the aquatic environment around Monaco. Marine Chemistry. 26: 119-132.

Imhoff, K.K. 1976. Taschenbuch der Stadtentwässerung. Oldenbourg Verlag. München.

Ivert, S. 1990. Bestimmung der Porenwasserzusammensetzung mit Hilfe der Dialysetechnik und frühdiagenetische Prozesse in Sedimenten der Eckernförder Bucht ('Hausgarten'). Diplomarbeit an der Christian-Albrechts-Universität zu Kiel.

Jahnke, R.A., Emerson, S.R., and Murray, J.W. 1982. A model of oxygen reduction, denitrification, and organic matter mineralization in marine sediments. Limnol. Oceanogr. 27(4): 610-623.

Jahnke, R.A., Emerson, S.R., Reimers, C.E., Schuffert, J., Ruttenberg, J., and Archer, D. 1989. Benthic recycling of biogenic debris in the eastern tropical Atlantic Ocean. Geochim. Cosmochim. Acta. 53:2947-2960.

Johnson, C.A., Sigg, L., and Lindauer, U. 1992. The chromium cycle in a seasonally anoxic lake. Limnol. Oceanogr. 37(2): 315-321.

Jørgensen, B.B. 1983. Processes at the sediment-water interface. in The Major Biogeochemical Cycles and their Interactions, edited by Bolin, B and Cook, R.B. John Wiley & Sons.

Jørgensen, B.B. and Sørensen, J. 1985. Seasonal cacles of O2, NO3- and SO42- reduction in estuarine sediments: the significance of an NO3- reduction maximum in spring. Mar. Ecol. Prog. Ser. 24:65-74.

Kabata-Pendias,.A. and Pendias H. 1984. Trace elements in Soils and Plants. CRC Press, Inc.

Kern, U and Westrich, B. (in press). Transport of sediment and associated heavy metals in a lock-regulated section of the River Neckar. Australian Journal of Marine & Freshwater Research.

Kersten, M., Förstner, U., Calmano. W., and Ahlf, W. 1985. Release of metals during oxidation of sludges - Environmental chemical aspects of dredged material disposal. Vom Wasser. 65: 21-35.

Kling, G.W., A.E: Giblin, B. Fry, and B.J. Peterson. 1991. The role of seasonal turnover in lake alkalinity dynamics. Limnol. Oceanogr. 36(1): 106-122.

Klump, J.V. and Martens, C.S. 1987. Biogeochemical cycling in an organic-rich coastal marine basin. 5. Sedimentary nitrogen and phosphorus budgets based upon kinetic models, mass balances, and the stoichiometry of nutrient regeneration. Geochim. Cosmochim. Acta. 51: 1161-1173.

Krom, M.D. and Berner, R.A. 1980. Adsorption of phosphate in anoxic marine sediments. Limnol. Oceanogr. 25(5): 797-806.

Krom, M.D. and Berner, R.A. 1981. The diagenesis of phosphorus in a nearshore marine sediment. Geochim. Cosmochim. Acta. 45:207-216.

Kuivila, K.M. and Murray, J.W. 1984. Organic matter diagenesis in freshwater sediments: The alkalinity and total CO2 balance and methane production in the sediments of Lake Washington. Limnol. Oceanogr. 29(6): 1218-1230.

Kuivila, K.M., Murray, J.W., and Devol, A.H. 1989. Methane production, sulfate reduction and competition for substrates in the sediments of Lake Washington. Geochim. Cosmochim. Acta. 53: 409-416.

Lee, B. and Fisher, N.S. 1992. Degradation and elemental release rates from phytoplankton debris and their geochemical implications. Limnol. Oceangr. 37(7): 1345-1360.

Lee, F.Y. and Kittrick, J.A. 1984a. Electron microprobe analysis of elements associated with zinc and copper in an oxidizing and an anaerobic soil environment. Soil Sci. Soc. AM. J. 48: 548-554.

Lee, F.Y. and Kittrick, J.A. 1984b. Elements associated with the cadmium phase in a harbor sediment as determined with the electron beam microprobe. J. Environ. Qual. 13(3): 337-340.

Li, Y. and Gregory, S. 1974. Diffusion of ions in sea water and in deep-sea sediments. Geochim. Cosmochim. Acta. 38: 703-714.

Manley, S.L., Goodwin, K., and North, W. 1992. Laboratory production of bromoform, methylene bromide, and methyl iodide by macroalage and distribution in nearshore southern California waters. Limnol. Oceangr. 37(8): 1652-1659.

Matisoff, G., Lindsay, A.H., Matis, S., Soster, F.M. 1980. Trace metal equilibria in Lake Erie sediments. J. Great Lake Res. 6:352-366.

Matisoff, G., Fisher, J.B., and McCall, P.L. 1981. Kinetics of nutrient and metal release from decomposing lake sediments. Geochim. Cosmochim. Acta. 45: 2333-2347.

Matsunaga, T., Karametaxas, G., von Gunten, H.R., and Lichtner, P.C. 1993. Redox chemistry of iron and manganese minerals in river-recharged aquifers: A model interpretation of a column experiment. Geochim. Cosmochim. Acta. 57: 1691-1704.

McCorkle, D.C. and Klinkhammer, G.P. 1990. Porewater cadmium geochemistry and the porewater cadmium: $\delta 13C$ relationship. Geochim. Cosmochim. Acta. 55: 161-168.

Moore, J.N., Ficklin, W.H., and Johns, C. 1988. Partitioning of arsenic and metals in reducing sulfidic sediments. Environ. Sci. Technol. 22: 432-437.

Morel, F.M.M. and Hering, J.G. 1993. Principles and Applications of aquatic chemistry. John Wiley & Sons, INC. 588 S.

Morfett, K., Davison, W., Hamilton-Taylor, J. 1988. Trace metal dynamics in a seasonally anoxic lake. Environ. Geol. Water Sci. 11(1): 107-114.

Morse, J.W. and Arakaki, T. 1993. Adsorption and coprecipitation of divalent metals with mackinawite (FeS). Geochim. Cosmochim. Acta. 3635-3640.

Mountfort, D.O. and Asher, R. 1981. Role of sulfate reduction versus methanogenesis in terminal carbon flow in polluted intertidal sediment of Waimea Inlet, Nelson, New Zealand. Appl. Environ. Microbiol. 42: 252-258.

Müller, D. and Schleichert, U. 1977. Release of oxygen-depleting and toxic substances from anaerobic sediments by whirling-up and aeration. In Interactions between sedimetns and Fresh Water. Edited by Golterman, H.L. The Hague: Junk Publ.

Müller, G. and Gastner, M. 1971. The 'Karbonat-Bombe', a simple device for the determination of the carbonate content in sediments, soils, and other materials. Neues Jahrb. Mineral. Monatsh. 10: 466-469.

Müller, G. Irion, G., and Förstner, U. 1972. Formation and diagenesis of inorganic Ca-Mg carbonates in the lacustrine environment. Naturwissenschaften. 59: 158-164.

Müller, G. and Prosi, F. 1977. Cadmium in Fischen des mittleren und unteren Neckars: Veränderungen seit 1973. Naturwissenschaften. 64: 530-532.

Müller, G. and Prosi, F. 1978. Verteilung von Zink, Kupfer und Cadmium in verschiedenen Organen von Plötzen (Rutilus rutilus L.) aus Neckar und Elsenz. Z. Naturforsch. 33c: 7-14.

Müller, G. 1980. Schwermetalle in Sedimenten des staugeregelten Neckars - Veränderungen seit 1974. Naturwissenschaften.67: 308-309.

Müller, G. 1981. Die Schwermetallbelastung der Sedimente des Neckars und seiner Nebenflüsse: Eine Bestandsaufnahme. Chemiker Zeitung. 105: 157-164.

Müller, G. 1986. Schwermetallbelastung der Sediment und Gewässergüte des Neckars 1972-1979-1985: ein Vergleich. in 2. Neckar-Umwelt-Symposium, 22-23.Oktober 1986 in Heidelberg. Heidelberger Geowiss. Abh. 5.

Müller, G., A. Yahya, P. Gentner. 1993. Die Schwermetallbelastung der Sedimente des Neckars und seiner Zuflüsse: Bestandsaufnahme 1990 und Vergleich mit früheren Untersuchungen. Heidelberger Geowiss. Abh. 69. 91 S.

Mun, A.I. and Bazilevich, Z.A. 1962. Distribution of bromine in lacustrine bottom muds. Geokhimya. 2: 199-205.

Murray, J.W. and Grundmanis, V. 1980. Oxygen consumption in pelagic marine sediments. Science. 209: 1527-1529.

Nembrini, G., Capobianco, A., Garcia, J., and Jacquet, J.M. 1982. Interaction between interstitial water and sediment in two cores of Lac Leman, Switzerland. Hydrobiologia. 92: 363-375.

Nriagu, J.O. and Dell, C.I. 1974. Diagenetic formation of iron phosphates in recent lake sediments. American Mineralogist. 59:934-946.

Norton, S.A., J.S. Kahl, A. Henriksen, and R.F. Wright. 1989. Buffering of pH depressions by sediments in streams and lakes. in Acid Precipitation, Vol. 4: Soils, Aquatic process, and Lake Acidification. Springer-Verlag.

Orem, W.H., Hatcher, P.G., Spiker, E.C., Szeverenyi, N.M., Maciel, G.E. 1986. Dissolved organic matter in anoxic waters from Mangrove Lake, Bermuda. Geochim. Cosmochim. Acta. 50: 609-618.

Park, S.W. and Huang, C.P. 1989. Chemical substitution reaction between Cu (II) and Hg (II) and hydrous CdS (s). Wat. Res. 23(12): 1527-1534.

Pedersen, T.F. and Price, N.B. 1982. The geochemistry of manganese carbonate in Panama Basin sediments. Geochim. Cosmochim. Acta. 46: 59-68.

Pedersen, T.F., Vogel, J.S., and Southon, J.R. 1986. Copper and manganese in hemipelagic sediments at 21°N, East Pacific Rise: Diagenesis contrasts. Geochim. Cosmochim. Acta. 50: 2019-2031.

Pettersson, K. 1986. The fractional composition of phosphorus in lake sediments of different characteristics. In Sediment and Water Interactions, edited by Sly, P.G. Springer-Verlag.

Pettersson, K. and Boström, B. 1986. Phosphorus exchange between sediment and water in Lake Balaton. In Sediment and Water Interactions, edited by Sly, P.G. Springer-Verlag.

Postma, D. 1981. Formation of siderite and vivianite and the porewater composition of a recent bog sediment in Denmark. Chemical Geology. 31:225-244.

Reimers, C.E. and Smith, K.L.Jr. 1986. Reconciling measured and predicted fluxes of oxygen across the deep sea sediment-water interface. Limnol. Oceanogr. 31(2): 305-318.

Reinhard, D. and Förstner, U. 1976. Metallanreicherungen in Sedimentkernen aus Stauhaltungen des mittleren Neckars. Neues Jahrbuch für Geologie und Paläontologie. 5: 301-319.

Sagemann, J., Skowronek, F., Dahmke; A., and Schulz, H.D. 1994. Saisonale Variation des Nitratabbaus in intertidalen Sedimenten des 'Weser-Ästuars'. In: Geowissenschftliche Umweltforschung. Edited by Matschullat, J. and Müller, G. Springer Verlag.

Sakata, M. 1985. Diagenetic remobilization of manganese, iron, copper and lead in anoxic sediment of a freshwater pond. Water Res. 19(8): 1033-1038.

Salomons, W., de Rooij, N.M., Kerdijk, H., and Bril, J. 1987. Sediments as a source for contaminants? Hydrobiologia. 149: 13-30.

Santschi, P., Höherner, P., Benoit, G., and Brink, M.B. 1990. Chemical processes at sediment-water interface. Marine Chemistry. 30:269-315.

Schulz, H.D., Dahmke, A., Schinzel, U., Wallmann, K., and Zabel, M. 1994. Early diagenetic processes, fluxes, and reaction rates in sediments of the South Atlantic. Geochim. Cosmochim. Acta. 58: 2041-2060.

Schwedhelm, E., M. Vollmer, and M. Kersten. 1988. Determination of dissolved heavy metal gradients at the sediment-water interface by use of a diffusion-controlled sampler. Fresenius Z. Anal. Chem. 332: 756-763.

Schwedt, G. 1993. Richt- und Grenzwerte Wasser - Boden - Abfall - Chemikalien - Luft. Umwelt Magazin.

Shaw, J.F.H. and Prepas, E.E. 1990. Relationship between phosphorus in shallow sediments and in the trophogenic zone of seven Alberta Lakes. Wat. Res. 24: 551-556.

Shaw, T.J., Gieskes, J.M., and Jahnke, R.A: 1990. Early diagenesis in differing depositional environments: The response of transition metals in ore water. Geochim. Cosmochim. Acta. 54:1233-1246.

Sigg, L. 1986. Metal transfer mechanisms in lakes; the role of settling particles. in Chemical Processes in Lakes. edited by Werner Stumm. John Wiley & Sons.

Sigg, L., M. Stumm, and D. Kistler. 1987. Vertical transport of heavy metals by settling particles in Lake Zurich. Limnol. Oceanogr. 32(1): 112-130.

Sigg, L, C.A. Johnson, and A. Kuhn. 1991. Redox conditions and alkalinity generation in a seasonally anoxic lake (Lake Greifen). Marine Chemistry. 36: 9-26.

Søndergaard, M., Kristensen, P., and Jeppesen, E. 1992. Phosphorus release from resuspended sediment in the shallow and wind-exposed Lake Aresø, Denmark. Hydrobiologia. 228: 91-99.

Song, Y. and Müller, G. 1993. Freshwater sediments: sinks and sources of bromine. Naturwissenschaften. 80: 558-560.

Suess, E. 1979. Mineral phases formed in anoxic sediments by microbial decomposition of organic matter. Geochim. Cosmochim. Acta. 43:339-352.

Sundby, B., Gobeil, C., Silverberg, N., and Mucci, A. 1992. The phosphorus cycle in coastal marine sediments. Limnol. Oceanogr. 37(6): 1129-1145.

Tessier, A., Rapin, F., and Carignan, R. 1985. Trace metals in oxic lake sediments: possible adsorption onto iron oxyhydroxides. Geochim. Cosmochim. Acta. 49: 183-194.

van den Berg, C.M.G. 1984. Organic and inorganic speciation of copper in the Irish Sea. Marine Chemistry, 14: 201-212.

van den Berg, C.M.G. and Dharmvanij, S. 1984. Organic complexation of zink in estuarine interstitial and surface water samples. Limnol. Oceanogr. 29(5): 1025-1036.

Vuynovich, D. 1989. Geochemische Untersuchungen an Porenwässern im Bodensee-Sediment. Dissertation am Institut für Sedimentforschung, Heidelberg Universität.

Wallmann, K. 1990. Die Frühdiagenese und ihr Einfluß auf die Mobilität der Spurenelemente As, Cd, Co, Cu, Ni, Pb und Zn in Sediment- und Schwebstoff-Suspensionen. Dissertation. Universität Hamburg-Harburg. 195 S.

Wallmann, K. 1992a. Solubility and binding forms of trace metals in anaerobic sediments. Vom Wasser. 78: 1-20.

Wallmann, K. 1992b. Solubility of cadmium and cobalt in a post-oxic or sub-oxic sediment suspension. Hydrobiologia. 235/236: 611-622.

Westerlund, S.F.G., Anderson, L.G., Hall, P.O.G., Iverfeldt, A., van der Loeff, M.M., and Sundby, B. 1986. Benthic fluxes of cadmium, copper, nickel, zinc and lead in the coastal environment. Geochim. Cosmochim. Acta. 50: 1289-1296.

Williams, T.M. 1992. Diagenetic metals profiles in recent sediments of a Scottish Freshwater Loch. Environ. Geol. Water Sci. 20(2): 117-123.

Winfrey, M.R. and Zeikus, J.G. 1977. Effect of sulfate on carbon and electron flow during microbial methanogenesis in frechwater sediments. Appli. Environ. Microbiol. 33: 275-281.

Wu, Y., Lin, Y., Guo, T., Wang, L., and Zheng, Z. 1992. Mechanisms of phosphorus released from the sediment-water interface in Xiamen Bay, Fujian, China. Science of the Total Environment, Supplement 1992: 1087-1097.

Yan, N.D. and Mackie, G.L. 1991. Contribution of zooplankton to the total cadmium pool in canadian shield lakes varying in acidity. Water, Air and Soil Pollution. 57-58: 635-644.

Printing: Weihert-Druck GmbH, Darmstadt
Binding: Buchbinderei Schäffer, Grünstadt

Lecture Notes in Earth Sciences

For information about Vols. 1–19
please contact your bookseller or Springer-Verlag